Green Solar Cities

The Green Solar Cities EU Concerto project focuses on the practical large-scale implementation of solar energy technologies in combination with new-build and retrofit low-energy building in the cities of Copenhagen, with its district Valby, in Denmark and Salzburg in Austria. This book aims to influence decision-makers in European cities towards a similar approach to the Green Solar Cities project, in close cooperation with leading building component suppliers, energy companies and engaged builders also working with local city officials.

This book will benefit those working towards 'Smart City' development and in need of clear policies on how to achieve this. In Denmark there are similar policies, with an overall aim to be CO_2 neutral by 2025 in the city of Copenhagen. However, there is still a lack of understanding concerning how solar energy as the world's number one energy source can play a major role here, and how this can be combined with energy-efficiency policies, use of district heating and combined heat and power.

The book introduces the international 'Active House' standard and work on 'Active Roofs' of the future. The connection between solar energy and low-energy building and energy renovation is to be ensured by use of the 'Active House' standard, which has been developed in cooperation with a number of leading building component manufacturers in Europe and Canada.

Green Solar Cities is published on the basis of the EU Concerto-funded project, Green Solar Cities, with Valby in Copenhagen, Denmark and Salzburg in Austria as the main city partners.

Peder Vejsig Pedersen is the Technical Coordinator and initiator of the Green Solar Cities project and has an MSc in engineering. Peder is also the Director of Cenergia and has been Chairman of the Danish Association of Sustainable Cities and Buildings (FBBB) since 2005.

Jakob Klint is an Administrative Coordinator of the Green Solar Cities project and Project Leader at the Kuben Management organisation, Denmark. Together with Peder Vejsig Pedersen he initiated the Valby PV plan in 2000.

Karin Kappel is an Architect MAA from Denmark and has been the Secretariat Leader of the Solar City Copenhagen association since 2004.

Katrine Vejsig Pedersen has an MSc in engineering. Her main focus is on sustainable organisation and planning in communities.

CONCERTO

Green Solar Cities

Green Solar Cities

Peder Vejsig Pedersen, Jakob Klint, Karin Kappel
and Katrine Vejsig Pedersen

Routledge
Taylor & Francis Group

LONDON AND NEW YORK

from Routledge

First published 2015
by Routledge
2 Park Square, Milton Park, Abingdon, Oxon OX14 4RN

and by Routledge
711 Third Avenue, New York, NY 10017

Routledge is an imprint of the Taylor & Francis Group, an informa business

British Library Cataloguing-in-Publication Data
A catalogue record for this book is available from the British Library

Library of Congress Cataloging in Publication Data
Green solar cities / [edited by] Peder Vejsig Pedersen, Jakob Klint, Karin Kappel,
and Katrine Vejsig Pedersen.
pages cm
1. Municipal engineering. 2. Solar energy. 3. Sustainable development. I. Pedersen,
Peder Vejsig, editor. II. Klint, Jakob, edtitor. III. Kappel, Karin, editor. IV. Pedersen,
Katrine Vejsig, editor.
TD159.G74 2015
696–dc23
2014021037

ISBN: 978-0-415-73119-5 (pbk)
ISBN: 978-1-315-77078-9 (ebk)

Typeset in Univers LT Pro 9/12 pt
by Fakenham Prepress Solutions, Fakenham, Norfolk NR21 8NN

Contents

Preface

Green Solar Cities is a policy for cities focusing on a holistic approach, which includes good, energy-efficient constructions and installations in buildings leading to comfort and a good indoor climate, combined with use of optimised energy-supply systems together with a local contribution from renewable energy sources. In this way it lives up to the aims of the EU Concerto programme, especially when good energy results are secured by use of an appropriate performance documentation as part of the strategy. This is already known to be a future demand of the EU Building Directive.

The Danish part of the Green Solar Cities EU Concerto project was defined on the basis of the agreed-upon Valby PV implementation plan from 2000, which aims at introducing at least 15 per cent PV solar power by 2025.

This PV implementation strategy spread from Valby, which is approximately 10 per cent of the size of Copenhagen, to the whole of Copenhagen from 2004 by the establishment of the public–private partnership Solar City Copenhagen.

The Valby PV plan features the involvement of local groups and organisations from Valby, mainly relying on continuous engagement from two of the authors of this book, Peder Vejsig Pedersen from the energy specialist company Cenergia and Jakob Klint from the Copenhagen Urban Renewal Company, which is now part of the Kuben Management organisation.

From 2004, when a PV implementation approach towards the whole of Copenhagen was decided, there has also been close cooperation with the third author, Karin Kappel, head of the Solar City Copenhagen Secretariat.

In order to be able to understand the background of the Green Solar Cities project it is relevant to look back to the start of the tradition of working with a combined focus on the use of solar energy and low-energy buildings, which dates back to the establishment of the Cenergia company in 1982 by four colleagues at the Thermal Insulation Laboratory, which was part of the Danish Technical University (DTU). It was here that the first zero-energy houses in the world were developed, primarily based on work by Professor Vagn Korsgård and civil engineer Torben Esbensen in 1973. This was good timing since it was the same year the first worldwide oil crisis emerged.

The first zero-energy house included a lot of innovation and many of the solutions did not work very well, but it led to numerous new research and technical development (RTD) initiatives in the 1970s and 1980s. A large number of practical implementation projects have been possible since then, together with a large number of engaged builders and cities, based on a combination of funding from the EU and national programmes in Denmark. In the period up to 2000 the main focus concerning solar energy solutions was on solar thermal systems and use of passive solar designs, coupled with a few innovative PV initiatives since the first EU-supported PV project in 1992.

As was the case with the development of wind turbines in Denmark, it has always been a bottom-up approach, with engaged local people and not strong governmental policies creating the basis for the development.

Around 2000 the first convincing solar power projects with PV emerged in Europe, with the most inspiring examples from Switzerland and the Netherlands. It was the city of Amersfoort in the Netherlands where 10 MWp of PV was installed in a new city area which led to the idea of establishing a PV plan for the Valby city part of Copenhagen. The plan was to reach 15 per cent electricity supply from PV by 2025, aiming at developing building- and infrastructure-integrated solutions with PV which would be accepted as a positive contribution to the city development. At the same time it was important to avoid a 'not in my backyard' development like in the wind turbine industry in Denmark, which led to a primary focus on more costly off-shore wind turbine parks.

With respect to the possibility of realising low-energy building and solar energy projects in buildings in Denmark, the main support since 2002 has been the EU Building Directive, which demanded a strong energy-saving policy in EU member states. The Danish Building Research Institute and the Danish Energy Agency made an open definition of low-energy buildings so the calculated energy frame value includes both solar thermal and PV systems, in order to comply with the demands of the EU Building Directive rules when they were introduced in the Danish Building Regulations in 2006.

In combination with an early agreement of the future demands for 2010 and 2015 (and later also 2020) in the Building Regulations and changes to the 'Danish Planning Law' which allowed municipalities to introduce future energy frame value-based building demands in their local planning rules, this meant a completely new situation for the building industry in Denmark, with a large number of low-energy class 2015 projects being realised, and at the same time also an important influence on large renovation projects as well.

But in the area of PV there was still a very limited implementation in Denmark until 2010; only 4 MWp was realised overall, and the existing net metering practice became a law. By 2011 the PV market in Denmark increased to 12 MWp and in 2012 due to continuously decreased PV costs, the market rocketed, with up to 500 MWp of PV installed. This led to a situation in which many people in single-family houses wouldn't have to pay any electricity costs for the next 20–30 years, a development which led to new solar PV legislation – by the end of 2012 the net metering was changed from yearly to hourly accounting in order to limit the market, also supporting a future-oriented smart-grid policy.

The situation in 2014 is unfortunately that the use of solar energy is still only seldom prioritised, even though many people accept that solar energy must play an important role as the basis of a sustainable energy supply in the future. Even in Denmark, where an 'Energy Agreement' from 2012 in the Danish Parliament aims at a 50 per cent wind energy contribution to electricity use by 2020, and a 100 per cent renewable energy source for electricity by 2035, there is still a reluctance towards using PV. And it is not accepted that PV systems, due to recent huge reductions in investment costs in many cases, can be more economical than off-shore wind turbine parks, even in the Danish climate with limited solar resources in the winter.

For the authors of this book there is no doubt about the need to develop a completely new practice for both new buildings as well as refurbishment of existing buildings, and there exists a large green-growth opportunity in Europe if a development of completely new types of active roofs and facades in the building industry and in European cities can be secured.

To give the best input to this development, it is our aim to showcase both positive and negative results from a large number of demonstration projects, including the realised Concerto demonstration projects in both Copenhagen and Salzburg.

In this book contributions have also been made by Vilfred Hvid (pp. 42–46), Inge Straßl (pp. 110–111) and Maria Wall (pp. 139–143).

Introduction

The Green Solar Cities Approach

The main aim of *Green Solar Cities* is to provide people with a vision of how cities and buildings of the future can be implemented with high energy quality, with an optimised energy supply which to a great extent is based on renewable energy, and with an equal focus on how to secure best practice by introducing a clear policy for performance documentation. This can be done by looking at experiences where the risks of bad performance results lie in practice.

The basic idea of this publication is to present examples from practice, including experiences from the EU Concerto project Green Solar Cities (2007–2013), to show that the idea of making low-energy buildings is about introducing a quality agenda for building. Without this approach, buildings and large renovation projects will be built in the traditional manner, which means they are leaky, full of thermal bridges and have a poor indoor comfort level. The passive house movement of the 1990s, mainly in Germany and Austria, demonstrated that high-quality constructions without thermal bridges and air leakages, combined with heat recovery of the ventilation air, resulted in buildings with nearly no need for heating.

When the passive house results were first discussed in Denmark after 2000, the ambition was to go in the opposite direction and mainly work with natural ventilation. Since many architects were used to this, the new passive house agenda was difficult to understand: 'Does that mean we will have to introduce mechanical ventilation again?'

This was actually the case. However, like introducing passive house qualities in construction, it was very important to communicate that a high-quality version of mechanical ventilation was needed. In other words, the efficiency of the heat recovery should be high and the electricity use for the fans should be low. However, due to a lack of clear standards for documenting these things in practice, it has unfortunately proved very difficult to control this. The experience is that it is very difficult to ensure that a mechanical ventilation system actually has low electricity use unless you can ensure a direct survey of this for the users and building owners.

This sets a good background for securing the next step in low-energy building; that is, buildings with easy access to all basic energy usage on a direct online basis. This is possible for different electrical and heating uses, as well as for basic comfort indicators like indoor temperature and humidity, CO_2 level and even daylight. The latest Wi-Fi technologies make it possible to obtain proof of an overall energy quality, documenting that users get what they have paid for.

This will constitute the basis of the final demand to create nearly zero-energy buildings, which includes the integration of renewables, especially the use of solar energy. Securing such a transformation is to a great extent a matter of integrating the solutions in an architectural way, so they benefit our cities and at the same time fit well into the existing energy supply systems.

This is what we might call the Green Solar Cities Approach.

If it is possible to secure the necessary documentation relating to energy and comfort, including use of renewable energy solutions, there will not only be a demand for this in Denmark – which has the world's highest energy costs – it will also appeal to countries that have no or very limited taxes on energy, like the UK and the USA. Demanding as much quality for buildings and cities as possible is natural, and it is clearly possible to document the performance so everybody can understand it.

Green solar cities with solar energy combined heat and power

The vision of *Green Solar Cities* is based on the universal relation between the necessary initiatives needed for the future. This means that energy savings in both new buildings and renovation should be introduced and optimised. An optimised energy supply solution and investigating how solar energy can be utilised with a high contribution in connection to this are important for making nearly zero-energy or positive-energy building possible.

When it comes to the market for low-energy building – both in new building projects and renovation projects – there are unique opportunities in Denmark. This is mainly due to the EU-based increased demands for new building energy standards in the EU Building Directives of 2002 and 2010, and the fact that an energy renovation strategy for existing buildings has been agreed in Denmark, aiming at renovating all buildings before 2050. The consequence of this is that the renovation rate, which today is around 1 per cent of the building stock per year, has to be increased to 3 per cent per year in order to create a completely CO_2-neutral society, which is the target to achieve by 2050.

There is, however, still an important barrier as low-energy building projects are still not realised based on holistic and high-quality solutions.

If we want to create green solar cities, one important example is the city of Copenhagen, which has changed its policy from not allowing solar energy solutions to be seen from public city and street areas to developing a real solar energy plan for Copenhagen. Best-practice solutions are supported and documented. Stakeholders are inspired by a solar energy city atlas showing where the best opportunities to utilise solar energy on buildings exist.

The opportunity to mix solar electricity with solar heating has not yet been implemented in Copenhagen on a large scale, but the EU-supported cooperation with the city of Salzburg in Austria over a seven-year period (2007–2014) has shown how city-based solar energy implementation, mainly based on solar thermal energy, can work in practice. For both cities it can be concluded that it is an obvious solution to mix the use of PV systems producing solar electricity with solar thermal installation in a 1:1 ratio so it matches the CHP-based district heating system. This kind of solar energy combined heat and power can be a realistic option for the future.

Outside the district heating areas the combination of continuously reduced PV costs and use of today's improved heat pump solutions also provides an interesting business case. A main obstacle is the lack of integrated performance documentation, even though continuously increased requirements concerning energy efficiency, like the low-energy classes for 2015 and 2020, are already in use in many cities. The result is that builders cannot be certain of obtaining the high energy performance that has been paid for, since the only documentation relies on calculations. Today, only one area of significantly increased energy quality requirements has been introduced in Denmark, and that is the demand for airtightness of buildings, since it is easy to control.

A clear challenge is to introduce full monitoring and survey systems that make it possible to establish the performance of a new building or a building renovation so this can be included in the way buildings are made.

In connection, it is very important not only to focus on how to make new building projects, but to try to utilise experiences from already realised solar low-energy building projects. Also, it is important not only to focus on the good results, but to learn from the not-so-good results, in order not to repeat the same mistakes again.

The big transformation towards the use of solar energy

Since the first oil crisis in 1973 it has been clear that solar energy, wind energy and other renewables are the energy sources of the future, since the amount of fossil fuel is very limited and CO_2 emissions are a serious problem.

At the same time it must be clear that one of the most important research and technical development areas is the combined focus on how to save on energy use, optimise the energy supply systems and combine the use of renewable energy sources with solar energy as the most promising option.

In the late 1970s there were only a few examples of solar energy utilisation. Since then a large number of solar low-energy new-builds and retrofit schemes have been realised, in many cases as either EU- or Danish-funded demonstration projects.

Since the new building rules were implemented in Denmark in 2006, based on demands from the EU, a large number of building projects have been realised with new, optimised constructions and installations have often been introduced in combination with the use of solar energy.

With the increased demands in the EU Building Directive of 2010 aiming at a nearly zero-energy standard for new-builds from 2018 in public buildings and from 2020 for all buildings, together with demands towards the EU member states to create a similar standard for existing buildings, it is now a challenge to create a whole new basis for energy-efficient building design in Europe.

There have been both positive and negative experiences during this period, based on which it can be stated that a main barrier of both energy-saving and solar energy solutions is that you cannot always be certain that building projects, realised in practice, will perform in accordance with calculations, which means performance verification is a very important task to focus on. In this book the main idea is to be open about problems and failures that have been experienced, showing from both older demonstration projects as well as recent projects a clear mix of both good and less good performance results; all this experience is relevant to learn from.

Chapter 1

Solar Energy in Cities

PEDER VEJSIG PEDERSEN AND JAKOB KLINT

Green Solar Cities: EU Concerto project

With the improved demands in the EU Building Directive from 2010 aiming at a nearly zero-energy standard for new public buildings by 2018 and all buildings by 2020, and with a demand towards the EU member states to create a similar standard for existing buildings, the challenge is to create a whole new basis of energy-efficient building design in Europe. At the same time the City of Copenhagen has an ambition to become the first carbon-neutral capital by 2025.

Extensive retrofitting of buildings, reorganisation of the energy supply and change in transport habits are some of many initiatives the City of Copenhagen will implement in order to become carbon neutral. With the Copenhagen Climate Plan the Danish capital combines growth, development and higher quality of life with a reduction in carbon emissions of around 1.16 million tons.

In the EU Concerto project, Green Solar Cities (2007–2013) (www. greensolarcities.com) EU funding has been utilised as a strong support for the large-scale PV implementation plan in the Valby part of Copenhagen. It was launched in 2000 and aimed at supplying 15 per cent of all electricity use in Valby using 30 MWp of PV electricity by 2025. By 2013 around 4 MWp of PV has been established, but only 600 m^2 of solar thermal installations. If this increased to 54,000 m^2 of solar thermal capacity, a true 1:1 solar energy combined heat and power solution will be available to document an optimised combination to the large-scale combined heat and power system in Copenhagen. This will actually utilise biomass in the future in the form of wood pills.

At the same time a number of new-build and housing renovation projects have improved their energy frame values by 30–79 per cent compared to normal practice.

Figures 1.1 and 1.2 show a gable with solar art directed towards the railway in Valby (by artist Anita Jørgensen). Here, PV is supplying electricity for neon light, which illuminates at night and is a landmark for the Valby PV plan.

In the Green Solar Cities EU Concerto project there is cooperation between the city of Salzburg in Austria and the local energy agency SIR. A

1.1
PV art at 'Prøvehallen' gable in Valby, which can be seen from the railway, is a symbol of the Valby PV plan.
Photo: Anders Sune Berg.

1.2
The PV gable in daylight. Photo: Anders Sune Berg.

so-called micro grid with district heating has been combined with 2000 m^2 of solar thermal solar collectors and a buffer tank in combination with a heat pump (Figures 1.3–1.5).

The idea is to introduce elements from the so-called 'Active House' concept (see www.activehouse.info) in relation to the Green Solar Cities project evaluation in Valby. A number of specifications are defined within the areas of *energy*, *indoor climate* and *environment*. Within *energy* focus is on *energy balance*, *energy design*, *energy supply*, *energy monitoring* and *verification and follow-up*. *Energy balance* is based on a calculation of all energy uses in a building, including electricity-using appliances and the energy supply system.

The Active House specification demands a procedure for energy monitoring, verification and follow-up. At present in Denmark focus is only on good calculation procedures, but there is no link between the calculations and the actual energy use in the building. This presents a good reason to introduce the same demands

1.3
Buffer tank for solar thermal energy in the
Lehen area of Salzburg.

1.4
New-build with solar collectors in Lehen, Salzburg.

1.5
Solar collectors and a buffer tank are part of a low-temperature micro grid. Photo: SIR (www.sir.at)

for 'verification' of all new-building projects within a two-year period, which has already been introduced in Sweden.

The 777 kWp (kilowatt-peak) PV installation at Damhusåen Waste Water Treatment Plant in Valby (Figure 1.8) cover an area of approximately 14,000 m² of secured landfill with a built-in liner below the grass. This land cannot be used for anything for many years due to pollution from residues of waste water. This example has huge prospects since waste water treatment plants represent 8 per cent of all electricity use in Denmark

ACTIVE HOUSE - Specification
Buildings that give more than they take

1st. Edition

1.6
Active House specifications.

1.7
The director of the 'Lynette Cooperative' Torben Knudsen was a driving force of the plan to realise the Damhusåen PV plant, here also supporting the large-scale PV plant for Valby.

1.8
Photo of the Damhusåen PV plant in Valby, which by January 2013 was the largest PV plant in the Nordic countries. It covers 8 per cent of the electricity at the Damhusåen waste water treatment plant, which is owned by the 'Lynette Cooperative' in Copenhagen (now Biofos). It is supplementing biogas-based electricity production to cover almost 50 per cent of yearly electricity use by renewables.

1.9
Photo from a large housing retrofit project in Valby at Hornemannsvænge housing estate: 14 kWp of PV (100 m²) and 100 m² of solar thermal is used for each of six renovated housing blocks.

At the Hornemannsvænge housing estate, low-energy retrofit solutions are used together with solar energy combined heat and power, where both solar thermal and PV electricity supplements energy from the large combined heat and power plants in Copenhagen.

Figure 1.10 shows a demonstration of PV-assisted ventilation in Valby in Copenhagen. The focus has been on documenting how low electricity use can be matched by PV electricity.

In 2000 the Valby PV plan was launched, introducing 30 MWp PV (300,000 m²), aiming for 15 per cent building-integrated PV for electricity use by 2025 in Valby in Copenhagen. An important goal to the PV implementation plan in Valby was to ensure that PV would be a positive element in the city development.

Peder Vejsig Pedersen and Jakob Klint

1.10
PV ventilation. Compact heat recovery ventilation for housing renovation where electricity use is matched by PV. A test of compact HRV unit from Ecovent/Øland is showing a dry heat recovery ventilation efficiency higher than 85 per cent.

1.11
An innovative building-integrated PV installation at the shared ownership housing cooperative 'Søpassagen' in Copenhagen. From a distance, it looks like a normal slate roof.

1.12
Close-up of PV art gable in Valby, a symbol of the Valby PV plan from 2000.

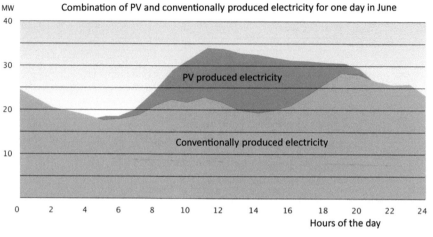

1.13
Brochure on solar energy in Valby funded by the EU Concerto project, Green Solar Cities.

1.14
An illustration of the electricity peak shaving in Valby when the PV plan is realised.

1.15
Development of a zero-energy test house in Denmark in 2003, which was exhibited in Valby.

1.16
Valby School in Copenhagen with building-integrated photovoltaic.

The Green Solar Cities project (www.greensolarcities.com) has, since 2007, included cooperation between Valby in Copenhagen and Salzburg in Austria. EU funding is utilised as a strong support for the large-scale PV implementation plan in Valby.

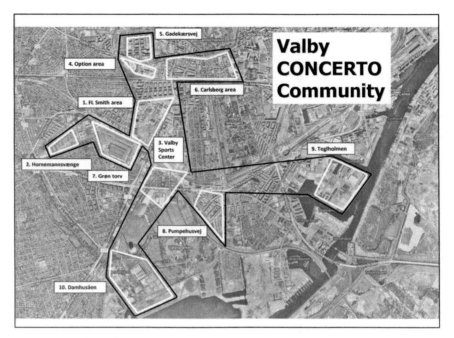

1.17
The EU Concerto area of Valby.

1.18
PV system at Sjælør S-train station in Valby.

1.19
PV balconies in Folehaven in Valby.

1.20
Thirty smaller PV projects were also supported by Concerto in Valby.

Peder Vejsig Pedersen and Jakob Klint

PV ventilation systems have been introduced in Valby and Copenhagen. This is compact *heat recovery ventilation* (HRV) for housing renovation where PV matches electricity use. Tests of compact HRV units from Ecovent/Øland show a dry heat recovery ventilation efficiency of greater than 85 per cent, alongside very low electricity use (Figure 1.10).

Table 1. 1 Expected development in PV costs in 2002: the target costs for 2020 were reached in 2012

Year	PV installation costs (DKK/Wp)	Electricity costs (ex. VAT) (DKK/kWh)	Pay-back time
2004	35–42	1.4	35 years
2009	30–36	1.7	23 years
2010	23–33	1.8	14 years
2015	18–27	2.2	9 years
2020	13–22	2.5	6 years

.Table 1.1 shows an indication from early 2011 of the historic and expected development of the economy for PV systems in Denmark up to 2020. In September 2012 the expected PV costs for 2020 had almost been achieved, and around 100 new PV installations – in particular for one-family houses – were made every day in Denmark, based on the net-metering scheme which leads to a pay-back time of around ten years.

US Department of Energy Secretary Steven Chu takes questions from the press after discussing the future of renewable energy in a plenary session of WREF. Photos (2): NREL

1.21
The US Department of Energy secretary at the WREF 2012 in Denver, proclaiming that Denmark had already reached grid parity for PV. This was unknown to the Danish government when it made its Energy Agreement the same year.

Solar is becoming cost-competitive

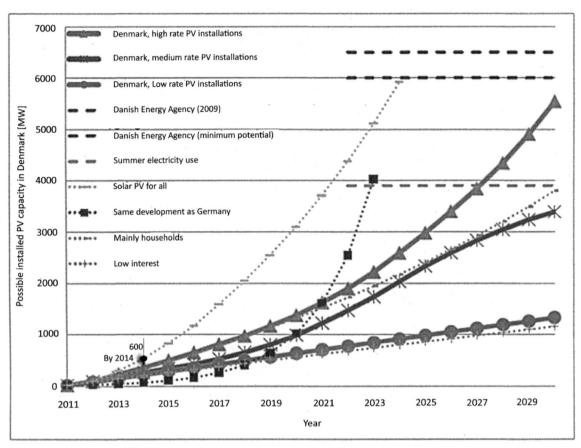

1.22

Possible development of PV in Denmark according to a basic and both more and less aggressive scenarios from 2009. In the basic scenario a PV capacity of 1 GW is obtained by 2020, and in 2030 the installed PV effect is 3.4 GW, equal to 8 per cent of the electricity used in Denmark. By 2014 600 MWp PV had been installed.

At the WREF 2012 (World Renewable Energy Forum) conference in Denver, the US minister of energy, Steven Chu, told the audience that at present PV systems were already cost competitive in Germany, France, Italy and Denmark, and that the same situation would be reached in the USA before 2020.

Example of demonstration projects with BIPV

The work on BIPV (building-integrated PV) demonstration projects in the Copenhagen area started in 1992 with the EU support for the 'PV in Valby' project. In 2000 a PV plan for an entire city part of Valby was launched in cooperation with the Urban Renewal Copenhagen company, the local electricity company, Copenhagen Energy and the municipality of Copenhagen. The Valby PV plan aimed at 15 per cent of electricity use coming from solar electricity by 2025.

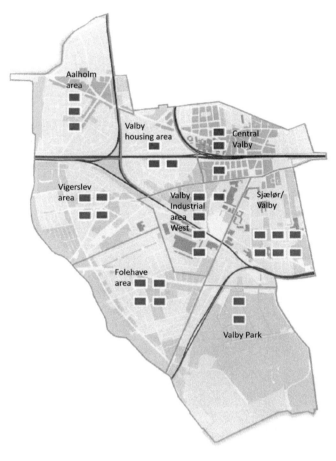

1.23
Valby PV plan aiming at 15 per cent solar electricity by 2025, including a lot of visualisation work, see also: http://greencities.eu/solivalby/index.htm

1.24
Brochure from 2000 on the PV plan for Valby, with the aim to cover 15 per cent of all electricity use by 2025 with 30 MWp of PV (approximately 300,000 m² PV).

Since then, work on practical PV implementation in Valby and in the rest of Copenhagen has been realised with support from EU projects like Resurgence and Concerto Green Solar Cities in Valby (www.greencities.eu; www.green solarcities.com), but also from the municipality of Copenhagen, which has given a total of approximately DKK12 million, and from the Danish Energy Agency as well as the PSO system in connection with the 'PV-Optitag project'. In 2004 the Solar City Copenhagen Association was established with support from the municipality of Copenhagen (www.solarcitycopenhagen.dk). In 2008 it advocated to kick-start use of BIPV due to the expected development in the economy of such systems. The PV activities in Copenhagen have also been promoted in relation to the member magazine of the Association of Sustainable Cities and Buildings in Denmark (www.fbbb.dk).

Here, it is relevant to look at the situation in Denmark, where it is being considered what to do with the combined heat and power-based district heating, when the main supply of electricity comes from wind turbines.

In relation to this, large-scale use of solar energy provides a win–win situation if it is installed in the form of so-called solar energy combined heat and power, which provides solar heating and solar electricity in a 1:1 ratio, in the same way as the existing large-scale combined heat and power system does. This means that operation of the large CHP system can be reduced in sunny periods, saving fuel as a result.

1.25
PV and Architecture, a book from 2005 showing best-practice BIPV examples.

1.26
An outline for Hedeparken housing association in Ballerup, with 1,000 apartments.

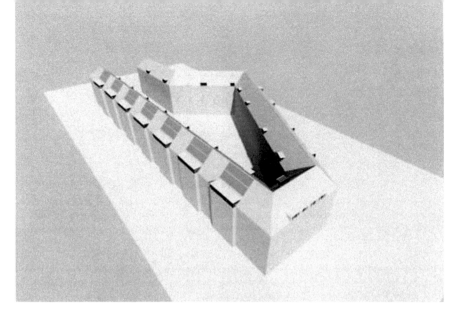

1.27
Example of proposed PV
integration with red-tile roofs
in Copenhagen.

1.28
Combined solar roof
(Canadian solar wall
preheating ventilation air) and
PV at Frederiksberg.

1.29
Zero-energy renovation in
2009, Albertslund in Denmark:
Solar Prism with PV mounted
on a flat roof.

1.30
Example of a PV project from early 2013 in Valby, Copenhagen. A new tile roof has been retrofitted with PV panels mounted on top. A more holistic – and cheaper – solution would be to use an underlay of asphalt-layer roof with PV modules on top, since red-tile roofs are very expensive.

1.31
Zero-energy housing area with 60 dwellings from 2012 in Tranbjerg Aarhus uses prefabricated Solar Prism systems from Danfoss/Velux with built-in heat pump, HRV system, PV panels and Velux roof windows.

An innovative PV installation at the shared ownership housing cooperative 'Søpassagen' in Copenhagen

By the end of 2011 a PV project with 45 kWp PV (approximately 360 m^2) for 90 apartments at Østerbro had been realised. This was the first BIPV installation in Copenhagen that was allowed by the municipality even though it can be seen from the street. The background for this new policy is the climate plan of Copenhagen from 2009, which aims at CO_2 neutrality by 2025.

Due to a location towards the 'city lakes' in Copenhagen and the very busy 'Fredensgade', the project was based on an intense and detailed dialogue with the chief architect's office in Copenhagen.

The PV system consists of 28 kWp PV on sloping roof areas and 17 kWp PV on the flat roof. The BIPV integration has been a success, although the cost for the sloping roof (DKK32,000/kWp) was more expensive (by approximately 50 per cent) than the flat roof installation (DKK21,000/kWp).

The design was developed by Solarplan in cooperation with Solar City Copenhagen, who supported an initial outline project. Energy and construction engineering were handled by Cenergia and MOE.

Another aspect of the project is the way the PV capacity beyond common electricity use was handled. The solution was to agree on a common purchase of all electricity by the tenants instead of the normal individual purchase. In this way yearly individual meter costs for each apartment were avoided and a better electricity price for a large consumer could be negotiated. Even with high costs for scaffolding and installation of a new meter system, a pay-back time of 13 years was obtained, with a positive balance for the tenants from the first year. At the same time it has been shown that for more simple installations on flat roofs a pay-back time as low as 6–8 years can be obtained based on this model.

1.32
Illustration from sketch design project for 'Søpassagen' at Østerbro in Copenhagen.

1.33
Detail of PV integration: slate roofing is maintained at the corner.

1.34
Nice integration of PV in an old 'Copenhagen roof'.

Peder Vejsig Pedersen and Jakob Klint

1.35
At the 'Søpassagen' shared housing project existing slate roofs were changed to PV. Very few people notice that this is a PV roof.

1.36
BIPV solution for 'Søpassagen' housing project in Copenhagen.

1.37
Integrated BIPV solution for the visible part of the roof, while the flat roof has an installation, which cannot be seen from the street area.

1.38
Example of guidelines from Copenhagen municipality has 'Søpassagen' on the front page.

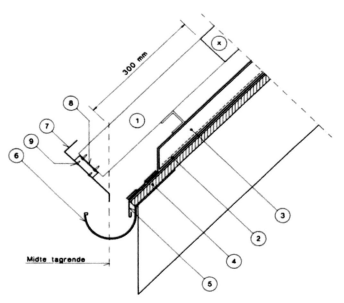

1.39
Detail of BPS-guidelines from Denmark on integration of solar thermal collectors on top of an asphalt-layer roof.

Peder Vejsig Pedersen and Jakob Klint

It is possible to use the previously developed BPS guideline for thermal solar collectors for PV integration in tile roofs.

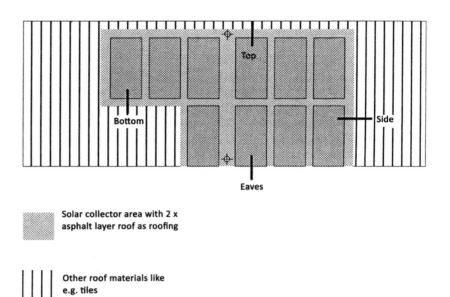

Top

Bottom

Side

Eaves

 Solar collector area with 2 x asphalt layer roof as roofing

Other roof materials like e.g. tiles

Solar collector modules

Flashing details

1.40
Illustration of solar collector area with other types of roof and solar collector modules, together with flashing details.

1.41
Example of PV integration. This was the winning project from the first PV architectural competition in Denmark. The project was realised in 1999.

Thirty years of solar energy development

The Green Solar Cities project is based on research and technical development (RTD) work concerning solar energy, which has taken place over a 30-year period. Back in the early 1980s solar energy primarily had to do with solar heating, both connected to solar thermal systems and also comprising passive solar heating and different kinds of solar wall systems.

Today, with the main focus on PV or photovoltaic solar electricity, it is worthwhile to include a focus also on solar thermal systems, which has led to interesting results in recent years. This is mainly in relation to large solar thermal systems connected to district heating, where in, for example, Denmark widespread use has occurred all over the country, with more than 500,000 m² of thermal solar collectors installed.

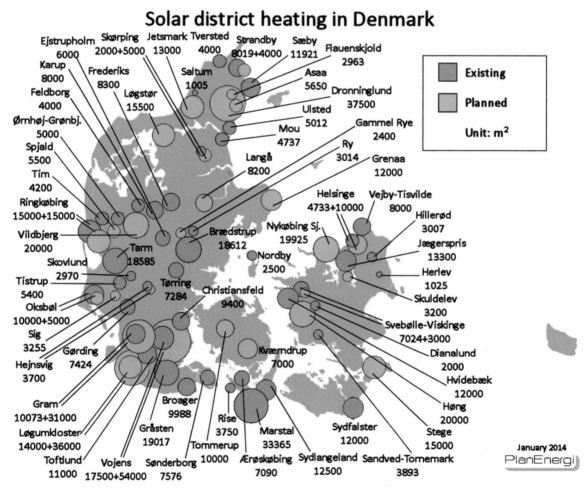

1.42
Overview of large solar thermal installations for district heating in Denmark by 2014.

Peder Vejsig Pedersen and Jakob Klint

1.43
33,300 m² solar thermal collectors in Marstal, Denmark, together with 75,000 m³ seasonal pit storage.

Recent examples comprise solar thermal systems in, for example, the smaller cities of Marstal and Dronninglund, which in combination with district heating and large seasonal storages in the ground, make it possible to cover 50 per cent of the yearly district heating demand with solar heating, which is remarkable in the Danish climate, which is cold and has only limited sunshine in the winter season.

At the city of Marstal on the island of Ærø with 1,500 inhabitants, an existing solar heating plant with 18,300 m² of solar collectors and 10,000 m³ of seasonal storage was in 2012 expanded with an extra 15,000 m² of solar collectors and a 75,000 m³ seasonal storage pit, together with a 1.5 MW heat pump in the EU-supported SUNSTORE4 project. This led to 55 per cent solar supply and 45 per cent biomass supply for the district heating in Marstal.

In the Tubberupvænge housing area in Herlev near Copenhagen in Denmark, the aim was to cover up to 70 per cent of the yearly heating demand for 100 low-energy housing units by solar heating. This was done by use of a combination of 1,400 m² primarily high-temperature, 12.5 m² large solar thermal collectors, together with a 3,000 m³ seasonal storage pit in the ground, working together with a heat pump, so it was possible to create temperatures up to 95 °C during summer and cool the bottom of the storage down to 5 °C in winter, which meant heat losses were limited to a reasonable amount on an annual basis.

The Tubberupvænge project was finalised in 1990 as the first solar heating project with seasonal storage in Denmark. The large solar thermal system in the Salzburg part of the Green Solar Cities EU Concerto project was made according to the same principles, based on design work from the Steinbeis Institute in Stuttgart in Germany, but with a 2,000 m³ buffer storage and a heat pump, together with 2,000 m² of solar collectors. With around 257 housing units, the targeted annual solar contribution is around 35 per cent. Recent projects in Marstal and Dronninglund in Denmark had seasonal storage sizes of 60,000–75,000 m³ and solar collector areas up to 50,000 m².

In the nearby city of Ballerup, also with funding from the EU, 700 m² of solar collectors were used for 100 prefabricated low-energy housing units, leading to a

1.44
Solar plant with seasonal storage and heat pump in Herlev, near Copenhagen.

Peder Vejsig Pedersen and Jakob Klint

successful result for the operation, together with low-temperature district heating and a local gas-fired CHP plant. This was the so-called Skotteparken project, which was not so successful with respect to the buildings themselves, where too high air leakages were observed, but nevertheless the holistic approach led to a world Habitat Award in 2004.

1.45
Optimised solar heating system for low-energy housing project Skotteparken in Ballerup, near Copenhagen.

1.46
Steel roofs and solar thermal collectors at Skotteparken.

Based on this project it was possible to form a European cooperation between housing associations, the European Housing Ecology Network and EHEN, and attract EU funding for large-scale ecobuilding demonstrations in seven EU countries, and at the same time realise a number of demonstration projects in the UK, where housing renovation of 350 apartments in the Stanhope Street housing project in Newcastle, in combination with Danish district heating solutions and use of a gas-driven CHP system, was the most interesting aspect.

In 1996 a European cities network, European Green Cities (www.european greencities.com), was established and got funding from the EU for demonstration projects in nine EU countries. The first examples of building-integrated PV systems were demonstrated in the Danish Urban Renewal demonstration project in the Hedebygade area in Copenhagen.

1.47
The heat recovery ventilation (HRV) units, integrated in the solar wall, shown behind the glazing at the Hedebygade housing block.

1.48
Urban renewal in the Hedebygade area in Copenhagen, with an example of a combined solar wall/ integrated heat recovery ventilation system operated by PV modules.

Cooperation with Salzburg and SIR was established here in connection with the realisation of a Salzburg region demonstration project in the city of Radstadt.

Other important RTD and demonstration projects from this period include a full-scale solar energy-oriented renovation of a research building at the EU Joint Research Centre in Ispra in Italy, which incorporated demonstration of a very cost-effective Canadian solar wall solution and very successful operation in practice documented by a detailed monitoring programme.

In Denmark at the same time it was possible to realise a number of EU-funded projects with the use of building-integrated PV solutions in the Lundebjerggaard housing project in the city of Ballerup as the most interesting. Here, RTD work was carried out on both PV solutions and ventilation designs as the basis of a large architectural competition, with an international jury led by NTNU from Norway. This led to good results in the end, although PV systems were still expensive and not economic for the users without some kind of extra funding.

1.49
Solar research building at JRC Ispra before renovation.

1.50
Solar research building at JRC Ispra after renovation with Canadian Solar Wall design.

By the end of the 1990s, PV demonstration initiatives included the Solar Village project in the city of Brædstrup in Jutland, and the Sunyard project in Kolding, with a BIPV demonstration for a large housing block.

In 1998 the rather small community working with PV implementation in Denmark visited the so-called PV city initiative in the city of Amersfoort in the Netherlands, which was demonstrating integration of 1 MWp PV in a new part of the city. This was a starting point for the city-oriented PV plan for the Valby city part of Copenhagen.

After 2000, when the PV plan for Valby was launched, the interest in use of PV increased in Denmark; there was a very useful BIPV development programme supported by the government, while the SOL-1000 initiative aimed at demonstrating the use of PV in single-family houses.

1.54
Example of a PV implementation project in London by the Peabody Trust, with supplemental funding from the EU-Resurgence project.

After 2001 a new government in Denmark with absolutely no interest in supporting renewable energy implementation was elected. At this point the EU-Resurgence project, together with supplemental funding from the City of Copenhagen and the SOL-1000 project, represented the only active implementation work with PV.

In particular, the funding opportunities from the City of Copenhagen were crucial, securing important BIPV demonstration projects including the SOLTAG project from 2005, with the CO_2-neutral rooftop apartment, which was realised with Velux as an important partner (www.soltag.net).

By 2007 the EU Concerto project, Green Solar Cities, started to focus on PV implementation in Valby in Copenhagen.

1.55
PV art project in Valby supporting the Valby PV plan.

Peder Vejsig Pedersen and Jakob Klint

1.56
The SOLTAG CO_2-neutral rooftop apartment, here exhibited at Velux's premises. Design: Architect maa Martin Rubow and Peder Vejsig Pedersen.

1.57
In Rønnebækhave II in Næstved, a small housing block was realised in 2006 as the first project in Denmark, based on the passive house principles and with a zero-energy heating design.

1.58
Green Solar Cities housing renovation in Valby with both PV and solar heating supporting the idea of solar energy combined heat and power. Here is the solar thermal part.

Copenhagen PV Co-op and reductions in PV costs

When the EU Concerto project, Green Solar Cities, was started in the summer of 2007, the Copenhagen PV Co-op had existed for three years and had 65 different investors who had invested in two PV installations of 40 kWp and 14 kWp, one on a municipal building and one on a building owned by the electricity company 'Dong Energy' (www.solcellelauget.dk).

This initiative was inspired by the Solar Stock Exchange model from Switzerland, which had good results for many years as a way to expand local investments in PV. During the realisation of the Green Solar Cities project it has been possible to create a PV project in Valby by Copenhagen PV Co-op with 22 kWp based on the same model.

The situation is that investments of this type will be difficult in the future since the model of selling PV electricity to people interested in supporting sustainable electricity production has been difficult to realise in Copenhagen, mainly due to very limited marketing of the concept. However, the fact is that the huge reduction in PV costs up to 2013 has changed the whole focus on PV implementation and the market as a whole. Figure 1.60 shows how the expected development of PV costs were in 2008, where typical installed costs was DKK35–38 per Wp (around €5/Wp).

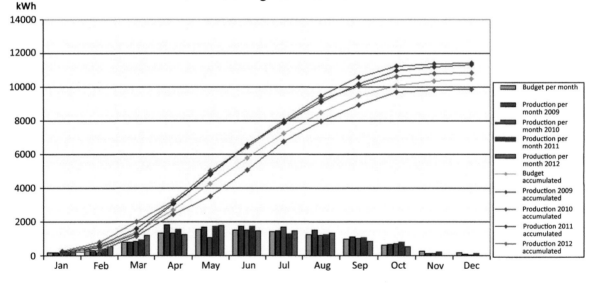

1.59
PV production and budget for Copenhagen PV Co-op. The Copenhagen PV Co-op
(www.solcellelauget.dk) has existed since 2004. These are PV production figures from the
Vigerslev PV system in Valby.

Peder Vejsig Pedersen and Jakob Klint

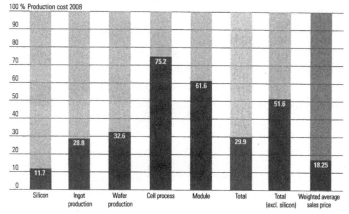

Reduction of production costs for multi c-Si modules between 2008 and 2012

100 % Production cost 2008

Silicon: 11.7
Ingot production: 28.8
Wafer production: 32.6
Cell process: 75.2
Module: 61.6
Total: 29.9
Total (excl. silicon): 51.6
Weighted average sales price: 18.25

1.60
Example of large cost reductions of PV production from 2008 to 2012. During the EU Concerto project from 2007 to 2013 the PV costs were reduced by a factor of five.
Photon International 12/2013.

Expected development in PV economy in Denmark (figures from 2008)

Table 1.2 Illustration of the development in PV cost reductions as predicted in 2004. In reality the cost aimed for in 2025 had already been achieved by 2014

Year	PV costs DKK/kWp	Expected electricity costs without VAT DKK/kWh	Pay-back time (years)
2004 establishment of Solar City Copenhagen	35–42	1.4	35
2009	30–36	1.7	23
2015	18–24	2.2	11
2025	10–12	3.0	4

By 2013 the targeted PV costs for 2025 had already been achieved in the best cases and for larger PV installations. Based on this a new situation emerged during 2012, in which grid parity was reached in Denmark due to the relative high electricity costs, and 500 MWp was installed compared to 12 MWp in 2011.

The big transformation towards use of solar energy

Since the first oil crisis in 1973, it has been clear that solar energy, wind energy and other renewables are the energy sources of the future, since the amount of fossil fuel is limited and CO_2 emissions are a serious problem.

At the same time it is clear, that one of the most important and needed research and technical development areas is the combined focus on how to save on energy use, optimise the energy supply systems and combine with use of renewable energy sources, with solar energy as the most promising option.

Since the new building rules were implemented in Denmark in 2006, based on demands from the EU, a large number of building projects with new, optimised constructions and installations have been introduced, often in combination with use of solar energy. With the improved demands in the EU Building Directive from 2010 aiming at a nearly zero-energy standard for new-builds from 2018 in public buildings and from 2020 for all buildings, together with demands towards the EU member states to create a similar standard for existing buildings, it is now a challenge to create a new basis of energy-efficient building design in Europe.

There have been both positive and negative experiences during this period, based on which it can be concluded that one of the main barriers of both energy saving and solar energy solutions is that it is not always certain that realised building projects will perform in accordance with calculations, which means performance verification is a very important task to focus on.

Table 1.3 World energy resources according to Richard Perez from SUNY Atmosphere Sciences Research Center, Albany, New York. In Denmark alone, a total of nearly 5 TW-year per year of solar energy is received every year, equal to one-third of the energy consumption of the entire planet

Renewable energies on Earth (TW-year per year)	Fossil fuels on Earth (TW-year)
Solar energy: 23,000	Coal: 900
Wind: 25–70	Uranium: 90–300
Biomass: 2–6	Petroleum: 240
Hydro: 4	Natural gas: 215
Geothermal: 2	
Tidal: 0.3	
Ocean thermal energy conversion: 3–11	
World energy use: 16	

The roadmap for PV from 2009 according to the International Energy Agency and EPIA (www.iea-pvps.org)

PV only constituted 0.1 per cent of global electricity generation in 2009, but was expanding rapidly due to dramatic cost reductions, with 40 per cent reduction in system prices from 2008 to 2009. Since 2000, with an average annual growth rate of 40 per cent, a new IEA Renewable Energy Division Roadmap envisions that PV will provide 5 per cent of global electricity consumption (900 GW PV) in 2030, with an annual market growth rate of 17 per cent, increasing to 11 per cent in 2050 (4500 TWh per year), corresponding to 3,000 GW installed capacity. European Solar Europe Industry (EPIA) projects in Europe alone 400 GW PV by 2020 and 700 GW by 2030.

1.61
Example of a housing area
with courtyard houses with
plans to be renovated with
solar energy combined heat
and power (PV + solar thermal
in 1:1 ratio).

Average efficiency of new module types continues to go up

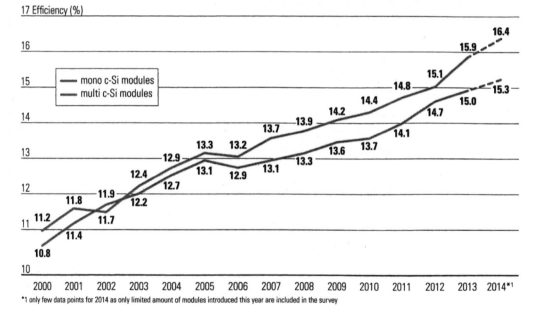

*1 only few data points for 2014 as only limited amount of modules introduced this year are included in the survey

1.62
The large increase of 46 per cent in PV production efficiencies from 2000 to 2014. *Photon International* 2/2014.

PV economy compared to off-shore wind turbine economy

The production cost for electricity from PV was, by October 2013, somewhere
between the cost for land-based wind turbines and off-shore wind turbines. This
is based on a PV installation cost of DKK14,000 per kWp for larger systems,
leading to PV electricity production costs of €90 per MWh.

At the same time it is a fact that the Danish energy company Dong Energy states that electricity from off-shore wind turbines have a production cost of €160 per MWh, with an aim to reach €100 per MWh by 2020. Compared to this it can be mentioned that large solar thermal installations like the Strandby heating plant in Denmark, with 8000 m² of solar collectors, produces solar heat for district heating at a cost of around €30 per MWh.

In Denmark there is an example of a large-scale solar heating supply for a whole city, Marstal, where a 55 per cent solar supply for district heating was secured in 2012, based on 35,000 m² of solar collectors, 75,000 m³ of seasonal storage and a heat pump, supplementing 45 per cent biomass-based heat.

Example of zero energy school design: the Haslev School in Denmark

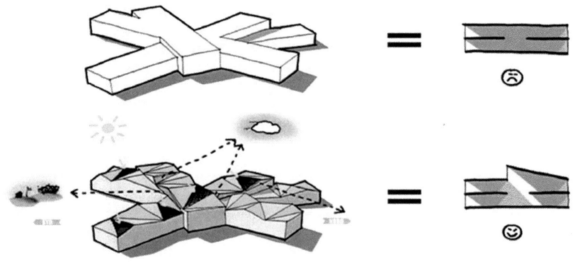

1.63
Example of low-energy class 2015 school with BIPV in Haslev, Denmark. The roof is designed according to demands for daylight and energy.

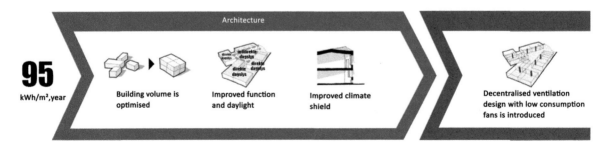

1.64
Design principles of a low-energy class 2015 school. First, the building volume is optimised with an improved function of daylight access and climate shield. Finally, a decentralised ventilation design with low-consumption fans is introduced. Arkitema Architects and Cenergia.

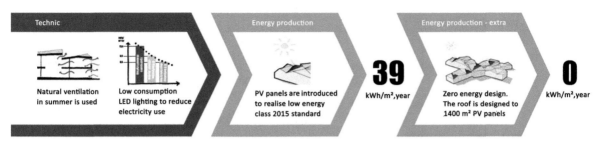

Technic	Energy production	Energy production - extra
Natural ventilation in summer is used — Low consumption LED lighting to reduce electricity use	PV panels are introduced to realise low energy class 2015 standard — **39** kWh/m²,year	Zero energy design. The roof is designed to 1400 m² PV panels — **0** kWh/m²,year

1.65

The vision was to reach a zero-energy design for the school in Haslev. In summer natural ventilation is used and low-consumption LED lighting is also used to reduce electricity use. PV panels are introduced to realise a low-energy class 2015 standard. With more PV panels even a zero-energy design could be obtained. However, this was not done in the end due to new PV legislation with hourly net-metering. Arkitema Architects and Cenergia.

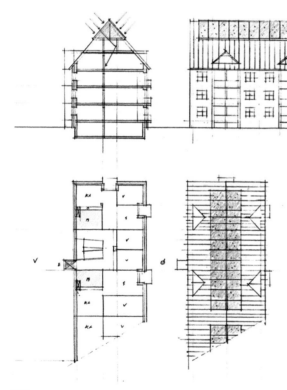

1.66
Possible integration of solar thermal and PV in an old housing block. Architect: Martin Rubow.

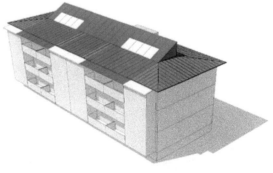

1.67
Example of housing renovation possibilities, based on solar energy combined heat and power, which utilises both BIPV and solar thermal collectors. Architect: Martin Rubow.

Samsø, 100 per cent renewable energy and sustainable energy island in Denmark.

1.68
The Danish island of Samsø has obtained a 100 per cent renewable energy supply mainly based on wind turbines. Here is an example of combined solar- and biomass-based district heating plant at Nordby.

1.69
At Samsø the 'Energy Academy' is disseminating knowledge and experiences concerning the 100 per cent renewable energy island. Now the work is focused on also being CO_2 neutral by 2020.

Peder Vejsig Pedersen and Jakob Klint

1.70
Close-up of the roof with
PV modules at the 'Energy
Academy' at Samsø. Architect:
Arkitema.

Danish building tradition and the road to low-energy buildings

The brick masonry tradition

The Danish building tradition differs from other Nordic countries, where wooden construction has been much more widespread. The access to clay and limestone has been limited at the Scandinavian Peninsula due to the geology. Wood has generally been in short supply in Denmark and had to be imported, and the few Danish forest areas have since early 1500 been reserved for naval shipbuilding. The Danish building tradition is similar to our southern neighbours like Germany and the Netherlands.

The Danish central city areas are dominated by brick masonry buildings. There has been plenty of clay, sand and limestone in the Danish earth to provide building materials such as brick and lime mortar, and the central city areas are dominated by brick buildings and blank masonry. Mud has, of course, been common as a building material as well, especially in the rural areas, but few buildings have survived.

From the early days of industrialisation the Danish cities were built with bricks – factories as well as residential buildings. Bricks were the rugged building material for solid construction and dominated the Danish construction industry and craftsmanship well into the 1960s, when concrete took over. A particularly significant period of brick constructions was 1900–1950, when the social housing associations dominated residential building construction in all cities in Denmark. In the period residential areas of very high quality and aesthetic value were built, with plenty of daylight, air and green open spaces compared to the residential areas in the dense inner cities. The construction of social housing areas peaked in the 1970s with the very large mono-functional concrete settlement areas on the peripheries of the cities.

Even today brick building is very significant in the Danish cities, as neither the Second World War nor the urban renewal policy have had a major impact on the older urban and residential areas. In the more dense parts of the cities the old blocks have been renewed by tearing down the buildings inside the blocks and crafting new open spaces and green courtyards. But seen from the street, nothing has changed and the renewal has not changed the image of the city.

The use of bricks has continued when constructing single-family houses, which is the dominant type of accommodation in Denmark. Large, detached areas characterise the suburbs of any Danish town, but today brick is only a facade material for buildings, and is no longer included as part of the supporting structure, although the illusion of brick house is maintained.

Dependence on fossil fuels

Brick houses are poor at retaining heat and the temperature of the buildings in winter is subject to the availability of fuels. Many probably froze in the winter time in years past, but the increasing availability of coal, coke, oil and gas has compensated for the lack of insulation. There was a very limited focus on buildings' energy consumption, both in relation to the existing buildings stock and to the energy demands for new buildings. Central heating ensured the homes were warm. The use of energy was nothing to worry about, there was plenty of coal and oil, and it was cheap.

There was little focus in the Building Code and few public demands for thermal insulation and reduced energy consumption. The regulations in the Building Code had mostly been focused on ensuring the quality of the construction, durability and fire codes. Additionally, there was a focus on the residential quality and it grew in importance due to the labour movement's influence during the period 1900–2000. In that period the social housing movement grew out of the desire and demand for healthier homes, and municipalities also took responsibility for ensuring good homes for the many that sought to move from the countryside to the cities, or moved from the poorest dwellings in urban centres.

With the oil crisis in 1972 the agenda changed, with considerable focus on the energy consumption of buildings and energy supply to them. Denmark's dependence on imported fossil fuels and rising prices resulted in a number of initiatives to reduce dependence and increase energy efficiency in buildings.

Simultaneously with all the government initiatives emerged a large number

Peder Vejsig Pedersen and Jakob Klint

of popular movements and grass-roots initiatives to promote energy reduction and renewable energy. Some had their roots in the environmental movement; others were new initiatives focusing on specific energy issues: resistance to atomic power, solar energy, wind and renewables in general.

There were launched different programmes and campaigns for energy saving in existing buildings; for example, initiatives such as insulation, sealing of houses, more efficient energy use. In the Building Code demands were focused on energy efficiency, and initiatives to foster combined heat and power were taken up and special schemes for financing district heating in urban areas were introduced.

It was particularly the following three points that influenced the use of energy in the housing and construction sector:

1. Combined heat and power with district heating.
2. Campaigns, subsidies, taxes on energy and other incentives for energy savings.
3. Tighter energy regulation for new construction through the National Building Code.

The 'waste' heat from coal-fired power stations was increasingly used to heat buildings. District heating was expanded in almost all Danish cities. These initiatives reduced energy consumption quite substantially per square metre, but led only to a stabilisation of total energy consumption in the period 1970–1990 because the heated area grew over the same period due to more heated area per capita.

In that period a large proportion of the profitable energy savings in the existing building stock were done. Typically, those with a pay-back period of less than five years were exercised during the period. These investments were stimulated by the higher taxes on energy and the public support of energy-saving investments.

Low-energy buildings

In the last 20 years of the twentieth century, renovation and urban renewal of the older multi-storey buildings from before 1900 has been a focus, with large public funding schemes and a large proportion of buildings being renovated. The purpose was two-fold: to improve housing and city qualities and to stimulate employment during the 1980s recession. The urban renewal properties went through intensive renovation schemes – new kitchens, bathrooms, windows and central heating – but there was limited focus on energy savings other than the change from oil and coke stoves to central heating by the district heating supply.

During the same period there was an ongoing tightening of energy rules in the Building Regulations in relation to the new buildings, which resulted in moderate improvements for the new buildings. In the same period, Denmark increasingly obtained access to oil and especially natural gas from the North Sea, so there were plenty of energy resources. Even though the high taxes on energy were maintained, initiatives that would reduce the use of energy more generally were weak.

The major changes and new demands on energy consumption for buildings (energy performance) came with the implementation of the EU Directive 'European Energy Performance Directive in 2002 (EPD)', which was implemented in Denmark in 2006. It has led to significant changes to performance requirements for new buildings. These changes came at the same time as the United Nations COP 15 meeting in Copenhagen and together with the financial crisis it changed the agenda in Denmark to a focus on the low energy performance of buildings. The construction industry focused on renovation, and limitations on new development projects made low energy performance a competing factor among builders. The ability to build to low-energy standards was important. The new buildings were, from this time, significantly more energy efficient and the construction industry is now well on track to meet the requirements for 'nearly zero-energy buildings'.

In the same period, a number of experiments involving low-energy refurbishment of the existing building stock were started. In particular, good progress was made in the renovation of concrete buildings from the 1960s and 1970s. It is possible to reach a high energy standard, but funding seems to be the biggest challenge in terms of achieving energy performance, since the energy savings can't finance the investments in energy savings.

In this context, brick buildings seem to be an even larger challenge due to the external insulation of the building envelope threatening to destroy the built heritage, and internal insulation being difficult technically. Until now there has only been a few good results with brick buildings, and the solution lies not only in achieving energy savings but, equally, in obtaining an energy supply based on renewables.

The Danish ambitions are high because there is a national commitment to be independent of fossil fuels by 2050. With such a goal, 3 per cent of the existing building stock should be energy-renovated each year to reach all buildings by 2050. Denmark is heading in the right direction, but there is still a very long way to go.

URBANISATION AND URBAN CHALLENGES

VILFRED HVID AND JAKOB KLINT

In addition to delivering a wide range of new low-energy buildings in Valby, the Green Solar Cities project has focused on the retrofitting of the existing building stock to low-energy standards.

One of the most extensive low-energy renovations that has been realised during the project is the renewal of the public housing 'Hornemannsvænge'. In Hornemannsvænge six concrete apartment buildings from the 1960s and 1970s have gone through extensive renovation. The degradation of concrete facades and general attrition has made it necessary to renovate the concrete, which has enabled insulation of the entire building envelope with 200 mm mineral wool and replacement of all windows with energy-efficient windows. The EU Concerto project has supported the renewal with better ventilation systems, solar thermal collectors and photovoltaics, and the buildings now meet the energy requirements for new buildings. Hornemannsvænge is a good example of a low-energy refurbishment, where the need for a general building renovation has been combined with a far-reaching energy optimisation.

Through the Green Solar Cities project, a concept for energy renovation of brick buildings has been developed. The brick buildings built before 1950 constitute a big challenge, because the building envelope of bricks lasts a long time, and the need for maintenance is significantly lower than for concrete facades. The synergy between facade maintenance and insulation cannot be obtained. At the same time, the brick facades are beautiful and external insulation is not allowed due to cultural heritage and the aim for preservation of the built environment. External insulation can only be done at the courtyard side.

The economically most realistic measures for reducing brick buildings' energy consumption are limited to:

- better windows with low U-values;
- internal insulation where possible, which is typically in the storey partition towards the basement;
- spandrels under the windows;
- insulation of the roof; and
- insulation of freestanding gables, etc.

In addition, savings may be obtained through better technical installations, such as ventilation with heat recovery rather than exhaust ventilation, through renewal and optimisation of heating and electrical systems, as well as through behavioural changes due to documentation of the individual use of heat and hot water. Finally, when replacing the roofs, it is possible to consider integrated photovoltaic and solar thermal collectors.

During the EU Concerto project period there has been ongoing dialogue with many housing associations, private landlords and owner associations to demonstrate low-energy solutions when retrofitting residential brick buildings. In general it has been very difficult to demonstrate low-energy measures related to the building envelope as attractive investments. Private landlords have no interest in energy savings as it is the residents who pay for energy, and in cooperative and owner-occupied housing the financial crisis has limited the economic associations' financial flexibility due to depreciation of the housing value. The best solution is to replace the roof cladding with insulation of the entire roof in combination with installation of photovoltaics.

In the Green Solar Cities project area the Copenhagen Municipality chose to prioritise Valby as an area for regeneration, which made it possible to co-finance the building renewal. Simultaneously, the Ministry of Housing, Urban and Rural Affairs decided that urban renewal funds could be used for energy improvements, which previously have been reserved for improvements of the

1.71
Black PV modules mounted on top of a new, red-tile roof in Valby. Could preferably have been integrated on top of plywood instead.

building envelope and the establishment of a bathroom, but not energy efficiency. This created the basis of a direct dialogue with the private landlords in Valby in terms of stringent energy retrofit, and a smaller property was considered for energy improvements.

In the summer and autumn of 2012 a process of intensive building improvements combined with intensive energy improvements were planned and sketch projects were started. With support from the Building Owners Investment Fund and the Ministry of Housing, Urban and Rural Affairs, the process of renewal began with a study trip to Basel and Zurich to look at some of the most extreme low-energy renovations realised on older buildings. The properties are the same age as the typical Copenhagen properties, which are from the mid-1800s to early 1900s.

A series of workshops were conducted with the participation of experienced people in the field of urban development, construction, rental housing and building renewal, and the process resulted in a conceptual design for the renewal of the properties in Valby.

The starting point for the workshops was urban challenges and future demands for housing. The City of Copenhagen takes part in the competition among cities, where cities are obliged to maintain and strengthen their local, national and international market position. Urbanisation continues with renewed strength, and the cities are in constant motion to attract new inhabitants and new jobs. Therefore, cities are also forced to grow denser to accommodate economic and population growth.

The need for more dwellings increases accordingly, because more and more people settle in the cities. At the same time, fewer and fewer people are living in families, which demands more space for living areas and significant requirements relating to the size of dwellings. The per capita size for living has generally increased over the past several decades. Dwellings in buildings from 1850–1930 are often very small and have small rooms, small kitchens and bathrooms (if indeed they have bathrooms), and they do not meet the future expectations for indoor climate, daylight and energy consumption.

The renovation concept for the properties in Valby is based on the following:

1. Densification of the existing cities without losing the values of the existing historical buildings' architecture and special qualities.
2. Creation of healthier and better dwellings for people living in cities.
3. Reducing the energy consumption by buildings.
4. Raising the quality of urban spaces and urban environments by maintaining and strengthening new qualities.

The possible solutions to these requirements were literally found on the top and the backside of the building. The facades of a typical multi-storey building from 1850–1930, which characterises the large parts of Danish urban areas, are typically built of good building materials and have many architectural details facing the street, while the facades facing the inner courtyards are often built in cheaper bricks and may appear with concrete render. The building facade facing the street typical defines the urban spaces and the perception of the city space.

1.72
Illustration of a so-called 'Living in Light' design for an old multi-storey housing block in Valby, Copenhagen with glass towards the backyard and on top of the building a part of an extension.

SOLUTION MODEL

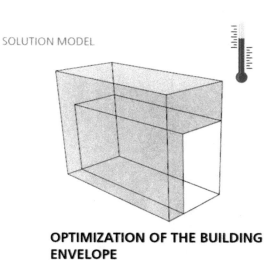

OPTIMIZATION OF THE BUILDING ENVELOPE

OPTIMIZATION OF THE HOUSING QUALITY - LARGER FLATS AND BETTER DAY LIGHT CONDITIONS

The transformation is relatively simple, embracing the building at the top and at the rear. It adds significantly more area to the building, creates better, bigger and healthier homes and a denser city, reduces building energy consumption to a level of new buildings and maintains the materiality of architecture and quality that are the hallmarks of urban spaces and urban environments.

Sketches and visualisations for the renewal of the properties in Valby have been prepared; the design and building process and refurbishment are expected to be ready in 2015.

Illustrations: living in light

1.73
Illustration of 'Living in Light' design towards the street area.

1.74
Illustration of a glazed extension towards the backyard.

1.75
An old housing block renovation in Valby according to the 'living in light' principles. At a meeting with Svendborg architects several roofing solutions which could utilise solar energy in the best way were discussed.

The Valby PV plan

Valby city district

Valby is situated in the southwest part of the City of Copenhagen and is famous for the hill 'Valby Bakke', where there is a view of the City of Copenhagen and the whole Øresund region. It is also famous as the 'home' of the well-known Carlsberg brewery.

Valby comprises 923 hectares, which is a little bit more than one-tenth of the area of Copenhagen. It is a mixed district with residential and industrial areas. The district of Valby is governed by Copenhagen City Council, but has its own local representatives.

The district has a population of 46,000 residents and 23,000 dwellings, which again is one-tenth that of Copenhagen. The district grew up during the industrialisation of Copenhagen and is now a typical urban area between the city and the new suburbs built after the war.

Residential and industrial areas

The residential areas consist of all types of buildings with respect to size and periods. Most of the urban development came after 1901, when Valby became part of the City of Copenhagen. The urban development took place until the end of the 1930s.

1.76
Valby is a district of
Copenhagen with 46,000
inhabitants.

The industrial areas house all kinds and sizes of trades and industries. There are 18,500 people working in the district. There is a wide variety of small enterprises performing all kinds of services and trades. Originally, larger enterprises predominated, but only a few remain and their activities have changed from manufacturing to administration and service. However, more manufacturing goes on in the district than in other parts of Copenhagen.

Valby is undergoing a transition from mixed industrial and residential use to a primarily residential area where several large former industrial sites are in the process of being rebuilt or used for other purposes. At the same time many of the existing residential buildings are older buildings, primarily from the 1920s–1960s and therefore not of an energy-efficient standard.

The Valby PV plan

In 2000 an ambitious PV implementation plan was set up for Valby. The aim of the plan is to supply a substantial part of the demand for electricity in the district of Valby via PV modules. These should be mounted on existing residential and

Peder Vejsig Pedersen and Jakob Klint

commercial buildings over a 25-year period (30 MWp/15 per cent electricity supplied by PV modules); in total this will be around 300,000 m² of PV modules.

The purpose of the PV plan is to gain technical, architectural and organisational experience in siting, establishing and operating a large PV system in an existing fully developed urban area.

This plan was formulated by the City Council of Valby, Copenhagen Energy (the utility company), City of Copenhagen, Cenergia Energy Consultants and Urban Renewal Copenhagen (see www.solivalby.dk).

The plan has now been in existence for 14 years and has led to a large number of PV implementations in Valby, and to the initiative Solar City Copenhagen (see www.solarcitycopenhagen.dk and the next chapter).

Chapter 2

Solar City Copenhagen

KARIN KAPPEL

2.1
Hestestaldskareen, København. In connection with the urban renewal of Vesterbro, one of the first solar cell systems in Copenhagen was installed in a former industrial building. Photo: Jens Lindhe.

Solar City Copenhagen

In 2004 a number of public and private players in the field of solar energy took the initiative in founding the organisation Solar City Copenhagen. The background to the initiative was an intention to make Copenhagen a demonstration and development centre for solar energy systems and energy-optimised buildings.

Moreover, the intention was for Copenhagen to stand out nationally and internationally and pave the way for business development in relation to reduction of buildings' energy consumption and integration of solar energy systems.

At the time, Copenhagen held a leading position in terms of CHP based on a well-developed district heating network. Wind turbines on Middelgrunden meant that a small amount of energy came from renewable sources, but, as regards solar energy and energy-optimised buildings, Copenhagen was lagging behind various other European cities.

Even though a number of development and demonstration projects of building-integrated solar energy systems had already been completed, increased measures were called for if Copenhagen was to stand out and be a development centre for the Øresund Region. The aim of Solar City Copenhagen's establishment was to contribute to such development.

The idea for the organisation was an offshoot of the cooperation in the Valby Solar Cell Plan, launched four years earlier as a large-scale demonstration project of solar cell integration in an existing urban district. The idea originated among the project participants, consisting of the City of Copenhagen, Valby Neighbourhood Council, Copenhagen Energy, Cenergia and Copenhagen Urban Renewal.

Inspiration also came from the Solar Cities Initiative, created by the International Solar Energy Society (ISES) as an international cooperation project between cities involved in solar energy. Such cities included Freiburg and Oxford, and, in 2004, the City Council of Copenhagen decided that Copenhagen should also be a member of the society and thus a Solar City.

2.2
A number of development and demonstration projects with building-integrated solar energy systems were completed, with subsidies from the Danish Energy Agency's SOL-1000 programme, the City of Copenhagen's RE pool or EU programmes, in 2000–2005. The Royal Danish Academy of Fine Arts, School of Architecture in Copenhagen thus had 63 m^2 of solar cells installed in the cafeteria's ridge light. In addition to contributing to the electricity used in the cafeteria, the solar cells also function as sun shading, and a ventilation solution powered by solar cells has been installed in the ridge light's gables as a supplement to ordinary building ventilation. Photo: Karin Kappel.

Karin Kappel

2.3
As a landmark for Valby Solar Cell Plan, this gable decoration was achieved
with 18 kW black solar panels and 150 mouth-blown fluorescent tubes, which
light up in the dark. Photo: Karin Kappel.

2.4
According to the artist, the gable decoration is an
image of a city map signalling a pulsating flow
of human activities and energy in the underlying
areas. The neon lights of the installation signal
city identity and say that this place is special
and that the city is close at hand. Photo: Anita
Jørgensen.

Establishment of the organisation

Solar City Copenhagen was established in the summer of 2004 as an organisation with bylaws and a board. The terms of reference set out the organisation's
overall objective, which was to contribute to a sustainable and carbon-neutral
energy supply through the development of solar energy systems, information
and communication of knowledge, as well as through regional and international
cooperation on solar energy projects. In all cases the focus is on architecturally
well-integrated solutions.

These objectives were to be achieved on the basis of cooperation
between players such as public and local institutions, private companies, building
owners, investors, manufacturers, consultants, research institutions and housing
associations.

Solar City Copenhagen secretariat

The City of Copenhagen funded a secretariat for running Solar City Copenhagen,
and also set up a renewable energy pool to support solar cell projects.

The secretariat was established with the intention for the organisation gradually to become self-financing and thus independent. This was achieved in the course of the following years, and in 2010 the secretariat moved from the City of Copenhagen's premises into Arkitekternes Hus (premises of the Danish Architects Association), which was also the point at which the organisation became nationwide.

Over time, the organisation has established an extensive network of firms of architects and engineers, local authorities, manufacturers, research and educational institutions, environmental organisations, energy companies and organisations such as the Danish Energy Agency, the Danish Technological Institute and the Architects Association of Denmark.

The network is both a national and international network and plays a significant role in the activities of the organisation, including the role of the secretariat as a hub for communication and as a knowledge bank. The network is also where the members of the organisation are found.

Activities

On an ongoing basis, Solar City Copenhagen completes a range of activities, all focusing on solar energy, energy-optimised buildings and energy renovation.

The organisation regularly organises theme conferences on specific and current problems in the area, often planned in cooperation with project partners or schools of architecture. Moreover, the organisation holds seminars at trade fairs on sustainable and energy-efficient buildings.

Other activities include technical field trips around Denmark and study tours to other European countries.

2.5
Study tour to Aarhus. Facade with thin-film PV. Photo: Lin Kappel.

Karin Kappel

2.6
A member of the Danish Parliament opens the European Solar Days at Solar City Copenhagen in Arkitekternes Hus. Photo: Karin Kappel.

2.7
Solar City Copenhagen conference. Photo: Karin Kappel.

Project participation

As part of its work, Solar City Copenhagen participates in a large number of projects; some of the current projects include:

- development of architecturally well-integrated building components with solar energy;
- development of less expensive roof or facade solutions with integrated solar energy;
- solar energy in industrial areas.

Solar City Copenhagen's principal role is communication and dissemination of the projects through conferences and publications, continuing education courses for architects as well as technical contributions to books and as a speaker. In addition to Danish projects, Solar City Copenhagen also takes part in international projects.

IEA SHC Task 41: solar energy and architecture

The aim of the project was to develop guidelines and show examples of high-quality architecture with active solar energy in an international perspective.

2.8
Case study: Energimidt. The office building has crystalline cells integrated in the facade and sliding glass shutters with thin-film PV. The facade composition gives the building a three-dimensional effect. Photo: Karin Kappel.

2.9
The website showing the case study collection.

From Danish quarters, Solar City Copenhagen contributed to the preparation of a website which offers a range of international examples. The results of IEA SHC Task 41 are available at http://task41.iea-shc.org.

The website also contains a communication guide for architects to be used as a tool for promoting solar energy solutions in building design in cooperation with customers and authorities. There is also a link to a website with a collection of products with integrated solar energy.

IEA SHC TASK 51: solar energy in urban planning

The objective is to develop strategies, methods/tools and guidelines, as well as to collect fine examples to facilitate increased use of solar energy in urban areas. The project focuses on improving and increasing the architectural quality of the integration of solar energy into urban planning so that the quality of the urban context is respected. The international project will be completed in 2017.

2.10
A competition project for a new urban district on the site of the Carlsberg brewery. The general principle is to combine energy consumption and energy generation so shops, business and housing directly use surpluses or losses of electricity and heat. The required energy supply will be based on photovoltaic, solar thermal and wind energy. There are solar cells on all the roofs of the urban district and tower houses with a thin-film facade, with a three-dimensional facade principle to optimise the energy-generating surface and achieve a dynamic expression, working as solar shading at the same time. The project will be one of the case studies in Task 51. Photo: Entasis Arkitekter.

Outline design

Solar City Copenhagen has an outline design scheme intended as an aid to overcoming the first barriers when a building owner is thinking about investing in solar energy. Questions such as the following are answered:

- Is the building suitable for solar energy in terms of architecture and structure?
- What are the possibilities of solar energy in the building and what solutions are recommended?
- What will the system look like architecturally?
- What will the system cost and what is its economy like?

The outline design is prepared by professional consultants and constitutes a decision-making basis on which the building owner can make his or her decision: is solar energy 'hot or not'?

Housing associations, companies, public institutions and multi-family homes can apply to the scheme, but single-family homes are not included in the scheme.

2.11
The residents used the outline design as a platform for roof renovation with a PV roof. Furthermore, the outline design could form the basis of applications for financial support. Photo: Karin Kappel.

Karin Kappel

2.12–2.13
Solar cells integrated into
the balcony glazing after
an outline design. Photos:
Arkitektfirmaet Ole Dreyer.

2.14
A new building in Albertslund
where the residents wish
to install photovoltaic. The
outline design proposes solar
cells in inclined glass slats
in front of the facade. Photo:
Johan Galster.

Solar energy integration

The use of solar cells on buildings in Denmark has developed from being an independent and added element on an already existing building into, to a higher degree, an element in integrated building design on a par with a number of other elements, but the purpose is still to reduce the need of the individual building for a supply of energy.

Previously, solar cells were often seen as nothing more than an energy-generating element which was installed without any aesthetic considerations over the interaction with the building itself. The large number of realised demonstration projects offer inspiration to be followed, and especially the Danish Energy Agency has played a significant role in this respect, as many national programmes, projects and grants over the past 15 years have focused also on the architectural integration into the Danish building culture.

The price of solar cells has plummeted over time. However, they are still expensive. The reason for this is that, today, solar cell systems are integrated into buildings to a much greater degree, and this generates more costs than merely the price of the individual panel.

Today, solar energy is a must-have for future-energy renovation facing the old building stock if both statutory requirements and environmental objectives are to be met.

In the local authorities, each individual system previously required building application processing, whereas today a large number of local authorities have issued guidelines for how to place solar cell systems, in order to stimulate building owners and ensure a successful architectural result. The City of Copenhagen and the local authority of Skive are two of the authorities working actively on photovoltaic.

2.15
Copenhagen University City Campus. A new building with photovoltaic integrated into the roof.
Photo: Erik Møller Arkitekter.

Karin Kappel

2.16
The roof is created with modules in different sizes, from 6 to 20 cells in one module. Photo: German Solar.

2.17–2.18
Roof renovation on an old property: the new roof has integrated solar panels. Photos: Dorthe Krogh.

Copenhagen

The City of Copenhagen has a target of 40 MW or 280,000 m² of photovoltaic by 2025 and plays an active role in terms of motivating people to install solar cells and supporting them in their efforts.

This is done through giving information to the citizens about the possibilities of establishing PV and a number of initiatives. One of the initiatives is the development of a so-called solar map which will give the citizens of Copenhagen

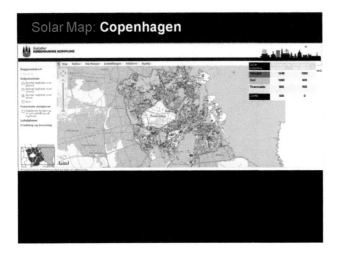

2.19
Solar map: the intention is to give house-owners an idea of whether a roof is suitable. It also delivers structured data extraction as three-dimensional city model contours and values of solar intensities. The tool can estimate potential in areas in terms of the amount of roof space (in square metres) within a given solar intensity. There are different views for different information, such as ownership, roofing materials, roof area size, SAVE values (how the level is for a protected building) and angles of roofs.

2.20
A new building with a flat roof and photovoltaic on a rack. Photo: Gaia Solar.

Karin Kappel

the first idea of whether their roofs are suitable for solar cells. The map shows selected solar irradiation ranges of the city's roofs. As it is key to the City of Copenhagen that photovoltaic in the city is used carefully in relation to its architecture and many preservation-worthy buildings, it is also possible to see the buildings' preservation value and the city's cultural environments and local plans.

The City of Copenhagen has adopted architectural guidelines for the installation of solar cells on the city roofs and also prepared a collection of examples with integrated solutions as a supplement to the guidelines. Further information is available at www.kk.dk (search for 'solceller').

Skive

The local authority of Skive aims at becoming energy self-sufficient by 2029 and has therefore made a vast investment in solar cell systems.

The project is named 'PhotoSkive' and comprises the installation of 1.4 MWp of solar cells on more than 70 local-authority buildings. The total investment amounts to DKK45 million (€6 million), of which DKK22 million was received from Energinet.dk through the ForskVE programme.

Specifically, the project has also tested different solar cell systems and demonstrated various ways in which to integrate solar cells into buildings so these can serve as an inspiration to others in the future.

2.22
Skive New Town Hall. Photo: Gaia Solar.

2.23
Solar shading with photovoltaic. Photo: Skive Municipality.

2.24
Skivehus School in Skive in Jutland is now a CO_2-neutral school based on PV and solar thermal installation. Photo: Energimidt.

One of the examples is Skive New Town Hall, which has solar cells integrated into the facade, on the glass shutters on the exterior solar shading, installed on the flat roof of the building. Combined, the 110 kWp solar cells generate 80,000 kWh annually.

The 4,200 m² building holds 165 workstations, and the energy consumption was originally calculated at 234,000 kWh. On account of the reduction in electricity consumption achieved thanks to LED lighting and ventilation systems with heat recovery, the annual electricity consumption is now only 156,700 kWh. Solar cells provide about 50 per cent of this amount and a mini CHP system generates the rest.

The building has succeeded in becoming Denmark's first plus-energy town hall thanks to 600 m² of solar cells, 162 m² of solar thermal collectors and a 25 kW biodiesel CHP system, which together generate 25 per cent more energy than consumed by the building.

Another example is Skivehus School, where rundown tile roofs have been replaced by an integrated solution with solar panels, which also function as solar shading. In this way, optimisation of the overall economy and aesthetics has been achieved.

The solar cell system was dimensioned to cover 80–85 per cent of energy consumption, and this allowed the school to reduce the remaining 15–20 per cent through energy management. The result is that in 2013 the school became completely self-sufficient in electricity.

2.25
The solar cells on the flat
roof of the kayak club have
been integrated into a flexible
material, which allows
people to walk on them
during cleaning. Photo: Skive
Municipality.

2.26
A number of measurements
were taken during the project,
here on vertically mounted
solar panels, which are also
used for hiding the technical
installations on the roof of the
building. Photo: Energimidt.

Karin Kappel

Chapter 3

Green Solar Cities

PEDER VEJSIG PEDERSEN

Green Solar Cities: Valby

The following is a presentation of some of the individual demonstration projects in the Concerto area in Valby and Copenhagen.

Henkel site II

Table 3.1
Henkel II (Building A)

Status	Finished
Gross floorage area/gross space, m²	6,840 m² (total area 13,149) office
Final energy consumption, kWh/m², year	Heat: 0 Electricity: 26.7 Total: 26.7
External wall	0.18–1.77 W/m²K (not all the envelope is being insulated)
Roof	0.1 W/m²K
Ground deck	0.2 W/m²K
Windows	1.5–1.7 W/m²K
Glazing	–

Two large 13,149 m² old industrial buildings have been renovated in connection with the EU Concerto project. One building was finished in September 2011 and the other in December 2011. The project includes extensive use of green roofs and use of a heat pump connected to air conditioning.

According to calculations, the energy frame value is around 69 kWh/m^2 per year. This is equal to a new 2010 building standard, which must be considered as good for an old renovated industrial building. The renovated industrial building, Henkel II, is now a public building.

A large heat pump system for heating and air-conditioning takes energy from the air and secures the total energy supply for heating and domestic hot water (DHW) through a DHW tank. Heating is supplied by radiators and through the air supply.

3.1
Henkel site II April 2012.

Peder Vejsig Pedersen

3.2
The industrial site was
originally owned by the
German Henkel industries.

3.3
Henkel site II October 2013.

3.4
Heat pump system from
Henkel II.

3.5
View to the interior of the renovated building.

Thermo photos

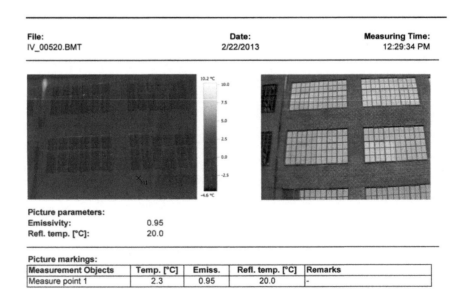

File:	Date:	Measuring Time:
IV_00520.BMT	2/22/2013	12:29:34 PM

Picture parameters:

Emissivity:	0.95
Refl. temp. [°C]:	20.0

Picture markings:

Measurement Objects	Temp. [°C]	Emiss.	Refl. temp. [°C]	Remarks
Measure point 1	2.3	0.95	20.0	-

3.6
Illustration of how most of the heat losses have been avoided besides a few spots beneath the windows.

Remarks:
Here it's possible to see a small heat loss underneath the window (spot M1). Also between the windows and underneath the other windows it's possible to see a small heat loss.

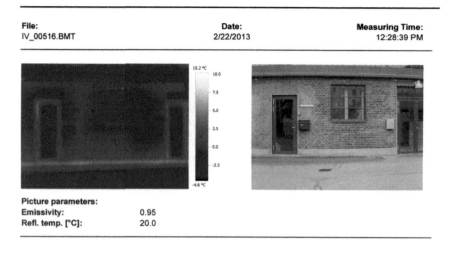

File: IV_00516.BMT	Date: 2/22/2013	Measuring Time: 12:28:39 PM

Picture parameters:
Emissivity: 0.95
Refl. temp. [°C]: 20.0

Remarks:
Here it's really visible that there is a heat loss in the foundation.

3.7
Here a heat loss in the foundation is very clear, shown by thermal photo.

Overall energy monitoring

The total electricity use is $47\,kWh/m^2$ per year, including both normal electricity use and heat pump electricity use, which is lower than the previous place the 'Bygningsstyrelsen' was situated. The total heating energy use is $62\,kWh/m^2$ per year.

Hornemannsvænge

Table 3.2
Hornemannsvænge

Status	Finalised
Dwellings	288 dwellings with six housing blocks
Gross floorage area/gross space, m^2	16,580 (total area 22,230 m^2)
External wall	0.14–0.18 W/m^2K
Roof	0.15 W/m^2K
Ground deck	0.4 W/m^2K
Windows	1.8 W/m^2K
Glazing	–
Final energy consumption, kWh/m^2, year	Heat: 52

3.8
Siteplan.

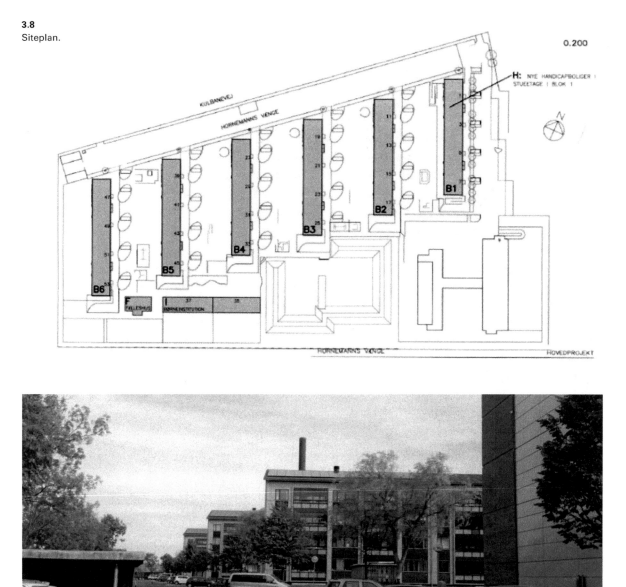

0.200

3.9
Hornemannsvænge in Valby
is a large concrete housing
retrofit project, with 288
apartments; this is the finished
renovation.

Renovation of the 288 dwellings in six social housing blocks was somewhat delayed, but completed by June 2013 – which made it possible to complete monitoring after implementation. A pilot project on ventilation systems with heat recovery was demonstrated in one apartment and the tenants approved the installation. An extended monitoring programme with 'before-and-after' monitoring has been approved as well. The results will be disseminated by AlmenNet, which is a network for sharing experiences on refurbishment of social

3.10
Before the renovation process.

3.11
Duct integration.

3.12
Traditional installation of PV. The photovoltaics that are used are monocrystalline. They are all oriented southwest with a slope of 30°.

3.13
PV production, including Block 4 until 16 October 2013, showed overall good results for nearly a year, with 7,558 hours of operation. The result was 14,012 kWh per year, equal to 983 kWh/kWp, which is good. A PV quality inspection by the Technological Institute showed overall good results of the PV installation.

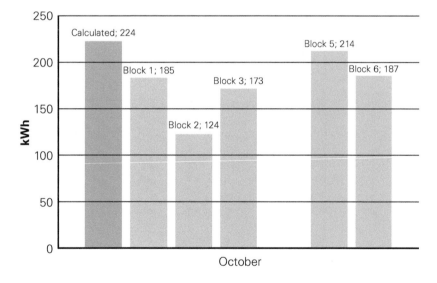

Electricity production in the 16–31 of October 2013

Calculated; 224
Block 1; 185
Block 2; 124
Block 3; 173
Block 5; 214
Block 6; 187

kWh

October

housing projects. Construction work started in August 2011 and new roofs with PV panels and solar thermal collectors have been established on all housing blocks, together with overall renovation works. Inspection of PV systems by the Technological Institute was successful. The monitoring system is connected to an overall energy management system (EMS), which was complicated to complete, so useful monitoring only exists from autumn 2013. The installed innovative solar energy combined heat and power system with both PV modules and solar thermal collectors (600 m^2 solar thermal and 85.5 kWp PV) has been monitored in detail.

Thermo photos

File: IV_00585.BMT	Date: 2/22/2013	Measuring Time: 2:01:29 PM

3.14
In the thermal photo you can see that most heat losses have been eliminated.

Picture parameters:
Emissivity: 0.95
Refl. temp. [°C]:   20.0

Remarks:
This picture is after the renovation, here you can see that there aren't the same spots of heat loss on the wall as before the renovation.

3.15
The old concrete housing blocks have been completely renovated with new facades and glazed balconies. Both PV and solar thermal collectors are integrated in the new tilted roof construction.

3.16
Integration of the ducts indoor.

3.17
Most of the glazed structure in the sunspace can be removed to avoid overheating in summer.

3.18
PV at Hornemannsvænge.
100 m² of PV modules were
made at a competitive cost.

3.19
100 m² of solar thermal
collectors for DHW were quite
costly (€800/m²).

Peder Vejsig Pedersen

Two types of solar energy systems, PV modules and solar thermal collectors, have been integrated in each end of the housing block at 2 × 100 m^2, working as a solar energy combined heat and power solution which matches the CHP-based district heating. In principle the district heating is not needed for domestic hot water in sunny periods in the summer.

A central heat recovery ventilation (HRV) system was chosen even though it is well known that the electricity use is higher than in decentralised ventilation systems and that the heat recovery efficiency is less efficient. The reason for

3.20
The new attic is insulated with paper granulate, which creates airtightness.

3.21
The central HRV system at Hornemannsvænge.

the choice of the central ventilation solution was to avoid access to apartments once or twice each year to change filters. With automatic filter boxes now being available, this should be possible to avoid in the future.

Monitoring of electricity use for centralised HRV systems in Hornemannsvænge documents a very high electricity use of 1,100 kWh per year per apartment. This is equal to a continuous electricity use of 125 W per apartment, which can be compared to monitored electricity use for good centralised HRV systems in new-builds of 55–60 W.

The difference is probably due to the more restricted possibilities for leading duct work through the building, since the engineering design seems to be of high quality. Such a high electricity use is also a result of the special fire demands given for centralised ventilation, so extra pressure losses of typically 100 Pa need to be introduced at armatures.

With installed HRV systems costing around DKK60,000 per apartment there is no economic reason to choose a centralised ventilation design, but it is still the most common solution for new-build and renovation housing, and it was very difficult in the Concerto area to convince any engineering companies to utilise designs with decentralised HRV systems.

Karens Minde

Table 3.3
Karens Minde

Status	Building finished May 2010
Dwellings	38 + 1 common house
Gross floorage area/gross space, m²	4,300 m²
External wall	0.21 W/m²K
Roof	0.12 W/m²K
Ceiling against unheated attic	–
Ceiling in cellar	–
Floor in cellar	–
Ground deck	0.12 W/m²K
Windows	1.4 W/m²K
Glazing	–
Final energy consumption, kWh/m², year	Type 00: Heat: 52.2 Electricity: 5.9 (Overheat: 4.4) Total: 58.1 Type 01: Heat: 49.7 Electricity: 4.6 (Overheat: 3.6) Total: 54.3 Type 02: Heat: 47.5 Electricity: 4.8 (Overheat: 3.6) Total: 52.3

Thirty-six newly built low-energy prefab dwellings were finished in May 2008 according to Danish low-energy class 2 (25 per cent better than actual energy requirements), with improvements to reach Concerto standard (30 per cent better). The total area of the dwellings is 4,300 m². A PV installation of 30.6 kWp was included in spring 2010. The dwellings of Karens Minde were the first dwellings of the affordable housing programme of Ms Ritt Bjerregaard, the former mayor of Copenhagen. The concept has now been taken over by a major social housing association, KAB, under the name Almen Bolig+ and spread out to the whole of Denmark in cooperation with other social housing organisations.

3.22
Outside view of 'Karens Minde' low-energy housing.

3.23
The project utilised prefabricated housing units from Latvia according to a special 'low-cost housing' plan.

3.24
Position of the photovoltaic.

3.25
Karens Minde. The picture shows the installed photovoltaics in detail. One panel is placed on the skylight (the boxes) and the other photovoltaics panels are placed directly on the roof in a montage system with a slope of 30°.

3.26
Karens Minde

Monitored electricity production for the whole photovoltaic plant

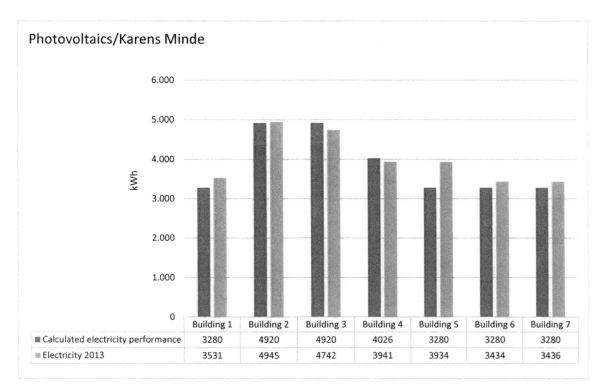

Photovoltaics/Karens Minde

	Building 1	Building 2	Building 3	Building 4	Building 5	Building 6	Building 7
■ Calculated electricity performance	3280	4920	4920	4026	3280	3280	3280
■ Electricity 2013	3531	4945	4742	3941	3934	3434	3436

3.27
The actual PV production was quite near to the calculated values.

Table 3.4
One of the inverters in building 5 (Inverter 22 D) was installed on 5 February, which to some extent can explain the good electricity production. In building 7 one of the inverters (Inverter 24 B) was first read on 9 February; here the overproduction is more realistic regarding the calculations.

	Building 1	Building 2	Building 3	Building 4	Building 5	Building 6	Building 7	Total
Calculated electricity performance [kWh]	3.280	4.920	4.920	4.026	3.280	3.280	3.280	26.988
Electricity 2013 [kWh]	3.531	4.945	4.742	3.941	3.941	3.934	3.436	27.964
Difference [%]	7.7	0.5	−3.6	−2.1	19.9	4.7	4.7	3.6

Conclusion from monitoring

The water consumption for all apartments (designed for three people) is 316.3 litres per day. That means 105.4 litres per person per day. The water consumption according to Building Regulations is expected to be 120 litres per person per day.

The district heating consumption is 65.2 kWh/m² per year (during the period 15 September 2008 to 18 September 2009). The calculated demand from Be06 is 51.1 kWh/m² per year. This means the consumption is 27.6 per cent higher than calculated. (Be06 was the first official calculation programme for energy in buildings, launched by the Building Research Institute, SBi, in 2006. Be10 was launched in 2010. It exists also in an English version and with climates outside Denmark.)

Electricity consumption in each apartment is 4,299.8 kWh/m² per year, which is 1,433.3 kWh per person per year. The average electricity consumption in Denmark is 1,423 kWh per person per year.

Final conclusion

Some parts of the prefabricated house construction weren't built to the required quality. When the windows and doors were installed some mistakes were made. But the demands for airtightness in the dwelling are fulfilled. There are still possibilities to save energy. As the indoor climate test showed, the temperature in the apartment is too high and can be reduced, therefore reducing heating consumption.

Langgadehus

3.28
One of many Green Solar Cities cycling tours, introducing solar low-energy buildings in Valby Copenhagen.

Table 3.5
Langgadehus

Status	Finished
Dwellings	Family housing: 59 Elderly housing: 68
Gross floorage area/gross space, m²	Family housing: 5,824 m² Elderly housing: 8,642 m² (including service centre)
External wall	0.14 W/m²K family housing 0.19 W/m²K elderly housing
Roof	0.10 W/m²K family housing 0.14 W/m²K elderly housing
Floor in cellar	0.28 W/m²K family housing 0.28 W/m²K elderly housing
Ground deck	– family housing 0.3 W/m²K elderly housing
Windows	Family housing 0.5 W/m²K (windows which cannot be opened) and 1.1 W/m²K (windows which can be opened) Elderly housing 1.5 W/m²K
Final energy consumption, kWh/m², year	Family housing: Heat: 25.4 (incl. solar collectors 15.7) Electricity: 7.1 Total: 32.5	Elderly housing: Heat: 86.7 Electricity: 6.2 Total: 92.9

3.29
Prefabricated family housing units from Latvia were used.

apartments on the ground floor and first floor. In addition to the senior centre there is housing on the second and third floor, comprising 59 family dwellings each of 100 m^2 (total 5,900 m^2), making the entire project 14,623 m^2. The building is supplied with 200 m^2 of solar heating. The work on the site started in autumn 2008 and the building was finished and ready for use and monitoring by March 2011. It took such a long time due to problems with very heavy rainfall, which damaged some of the apartments. The 59 prefabricated rooftop dwellings at Langgadehus are built according to the same balanced building principles as the Energibo concept (carbon-neutral rooftop concept).

3.30
Photo showing part of the 200 m^2 thermal solar collectors placed on the roof of the 'Langgadehus' building. The experience was that costs were 3–4 times higher per m^2 of solar collectors compared to installation of large areas on the ground.

3.31
Illustrations from Langgadehus. The super low-energy prefabricated family dwellings in two storeys were placed on top of a new on-site built nursing home for old people in Langgadehus in Valby. There is solar DHW heating from the 200 m^2 solar thermal system on the roof.

3.32
A large array with 200 m² of solar thermal collectors at Langgadehus in Valby. The PV panels in front on Valby School are cooled with a heat pump to make domestic hot water.

Energy supply

Energy supply is from the following sources:

- district heating;
- solar heating system (200 m²), with priority for the family dwellings but an export option to the senior housing;
- heat recovery ventilation (an innovative concept for family dwellings, including detailed monitoring).

3.33
District heating information.

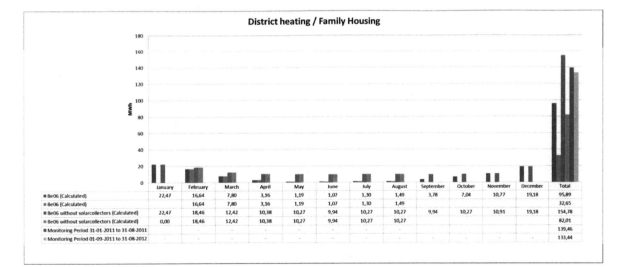

District heating / Family Housing

	January	February	March	April	May	June	July	August	September	October	November	December	Total
■ Be06 (Calculated)	22,47	16,64	7,80	3,16	1,19	1,07	1,30	1,49	3,78	7,04	10,77	19,18	95,89
■ Be06 (Calculated)		16,64	7,80	3,16	1,19	1,07	1,30	1,49					32,65
■ Be06 without solarcollectors (Calculated)	22,47	18,46	12,42	10,38	10,27	9,94	10,27	10,27	9,94	10,27	10,91	19,18	154,78
■ Be06 without solarcollectors (Calculated)	0,00	18,46	12,42	10,38	10,27	9,94	10,27	10,27					82,01
■ Monitoring Period 31-01-2011 to 31-08-2011	-	-	-	-	-	-	-	-	-	-	-	-	139,46
■ Monitoring Period 01-09-2011 to 31-08-2012	-	-	-	-	-	-	-	-	-	-	-	-	133,44

District heating for the family housing area (the basement is not included): there is a 70 per cent higher consumption than calculated. This is partly influenced by operational issues after the housing units were brought into use in February 2011. With a total monitored district heating use of 23 kWh/m^2 per year for the seven-month period, it is likely to be better than the Concerto aim of 46 kWh/m^2 per year.

For the period 1 September 2011 to 31 August 2012 the energy use is lower than the period before, even though this time the heating season is included. The reason for this might be the fact that the solar collectors are now contributing with solar energy, even though there are still problems getting the system to work properly. The district heating consumption is still higher than calculated at around 39 per cent.

District heating – space heating for the family housing area: for the period 31 January 2011 to 31 August 2011 the figures show that the dwellings used more energy compared to the calculations in the Be06 tool. A calculated room-heating demand of 5.5 kWh/m^2 per year seemed very low compared to realistic, in practice, energy consumption.

Some explanation can be found in the Be06 calculation, where the energy for heating relates to the presumption that the indoor temperature of the dwellings will be 20 °C, whereas, in general, people prefer to have a higher indoor temperature. If the temperature is raised to 22 °C in the Be06 calculation the energy consumption for the monitored period will go up to 20.6 MWh, which still is lower than the actual monitored consumption. But again, based on the 6,088 m^2 housing area, the monitored heating use in the seven-month period is still only 11.3 kWh/m^2 per year, which is actually very low.

In the next period, 1 September 2011 to 31 August 2012 the energy used for heating the apartments was reduced, but still higher than the calculation. This is the same conclusion as in many other building projects: it is not possible to rely entirely on the Be06 calculations when predicting the actual heating consumption.

The actual energy use for space heating is 10.2 kWh/m^2 per year for this year, which must be considered good and much better than the Concerto expectations; however, it is still higher than the ideal Be06 calculation of 5.5 kWh/m^2 per year.

For the period 31 January 2011 to 31 August 2011 the hot water consumption calculated in Be06 and monitored is almost the same (Be06 = 884.2 m^3; monitored = 882.0 m^3). The monitoring shows that the building that covers the family housing is using 54 MWh for heating the hot water. The Be06 calculation shows that the building should use between 23 MWh (with solar collectors) and 70 MWh (without solar collectors), which means there is a good connection between the actual district heat consumption and the calculated district heating consumption.

For the period 1 September 2011 to 31 August 2012 the energy consumption for the hot water is almost the same as the Be06 calculation with solar collectors. Considering that the hot water consumption has been monitored to be 27 per cent higher than calculated in the Be06 calculation for the same period, it is notable that the energy use is so low.

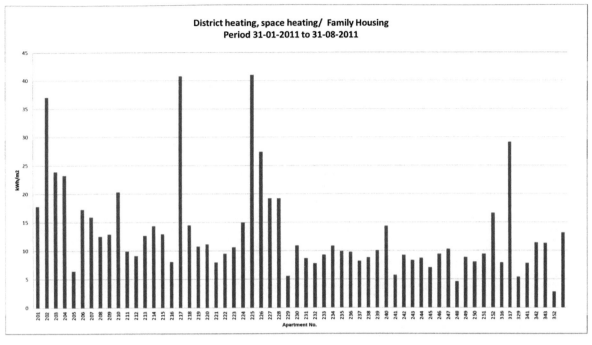

District heating, space heating/ Family Housing
Period 31-01-2011 to 31-08-2011

3.34
District heating consumption for the space heating for each apartment in the family housing part for the period 1 September 2011 to 31 August 2012. There is a high variation in district heating use, where some dwellings use five times more district heating than others.

The heat recovery is 85 per cent and the air change is 0.37 l/m/s.

District heating and space heating in elderly housing: the district heating consumption is 37.7 per cent higher than calculated in the Be06 calculation. In the Be06 calculation it is calculated that the energy for space heating is 65.1 kWh/m². The reason the heat demand is so high is the extra-high ventilation demand that 'institutions' have. The average air change in the building is 0.75 l/m/s, whereas it is 0.35 l/m/s for private dwellings according to the national Building Regulations. There is also a central ventilation solution that has a low heat recovery of around 65 per cent.

District heating, space heating, different units – senior housing: indoor climate values in two apartments have been monitored by the Danish Building Research Institute, SBi. The results document the indoor temperatures, relative humidity and the absolute ventilation rate to establish effects from demand-controlled ventilation. Overall, the indoor climate figures are acceptable for the users.

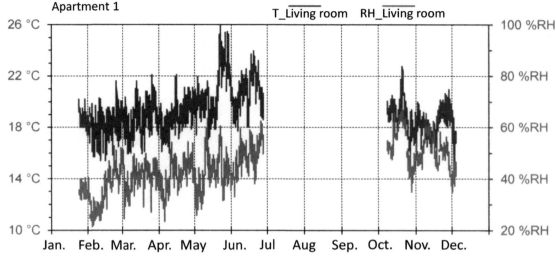

3.35
Mean values of temperatures and relative humidity monitored in two places in the living room of apartment 1.

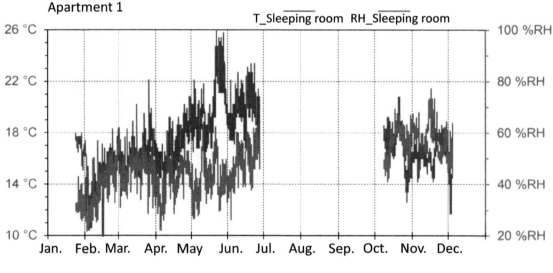

3.36
Temperature and relative humidity in the sleeping room in apartment 1.

Water Cultural House

This is a 3,230 m² new public swimming pool with service areas built according to low-energy class 2. Part of the electricity consumption is supplied by a 19.1 kWp PV system and heat is supplied by the district heating network. Building started in March 2010 and the official opening of the centre was on 15 March 2012.

The Water Cultural House is state-of-the-art in low-energy performance for indoor swimming facilities in terms of both building design and water treatment technology. The air and water temperature are at least 28 °C in order to achieve the best comfort for visitors. The building is now in full use and the monitoring programme started 1 May 2012.

Table 3.6
The Water Cultural House in Valby sports park

Status	Finished
Gross floorage area/gross space, m^2	3,230 m^2
External wall	0.15 W/m^2K
Roof	0.08 W/m^2K
Ceiling against unheated attic	–
Ceiling in cellar	–
Floor in cellar	0.11 W/m^2K
Ground deck	0.12 W/m^2K
Windows	1–1.5 W/m^2K
Glazing	–
Final energy consumption, kWh/m^2, year	Heat: 22.95 Electricity: 10.9 Total: 33.85

3.37
Outside view of the 'Water Cultural House' in Valby.

3.38
Ariel view of the 'Water Cultural House' in Valby, where the PV areas on the neighbouring building is also taken into account in the energy frame value calculation.

3.39
The Water Cultural House in Valby was developed with funding from the EU Concerto programme.

3.40
The foyer of the 'Water Cultural House' in Valby.

3.41
PV for the Water Cultural
House building is placed on
the neighbouring building, the
Valby Sports Arena.

Thermo photos

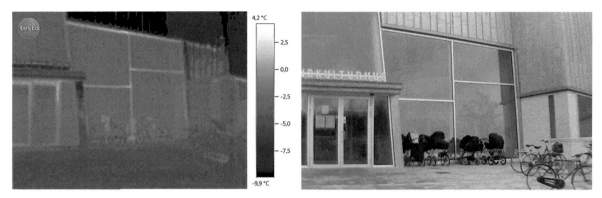

3.42
It is clearly shown that the frame has a higher heat loss than the glass, but this is normal. The reason the zinc is black is due to infrared reflection from the thermal camera.

It was acceptable according to Building Reg. calculations,to have an energy use of 70,5 kWh/m² per year, but the actual energy use is only 55,97 kWh/m² per year.

Yearly district heating use:
91,11 MWh

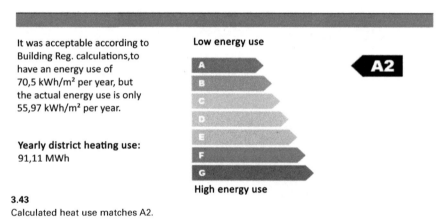

Low energy use

A

B

C

D

E

F

G

High energy use

A2

3.43
Calculated heat use matches A2.

Optimised facade design ...... U 0,8 - 1,4 W/m²°C

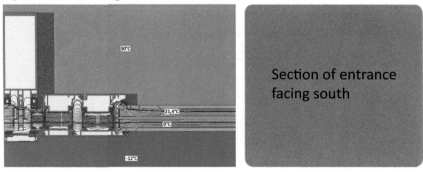

Section of entrance facing south

30°C

21,4°C

0°C

-12°C

3.44
Calculated heat loss of the glazed area at the entrance, with U-values between 0.8 and 1.4 W/m²K.

Peder Vejsig Pedersen

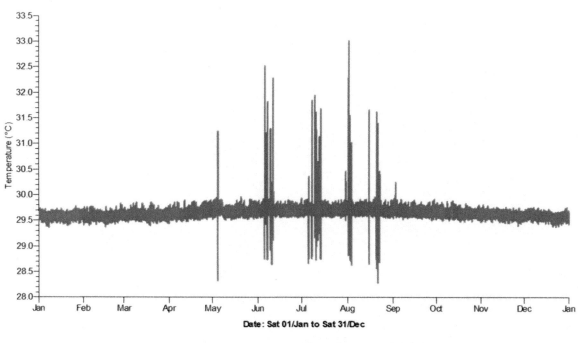

3.45
Air temperature in the swimming hall calculated for the whole year. The mean level is high, at around 29.5 °C.

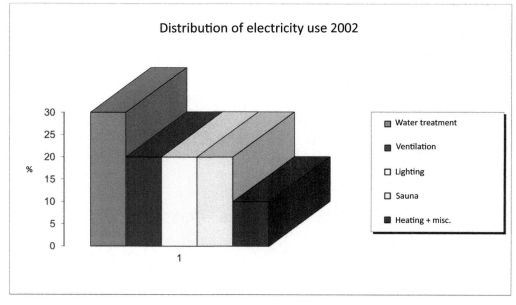

3.46
Distribution of electricity use (percentage of total).

50 m pool, electricity use for pumps

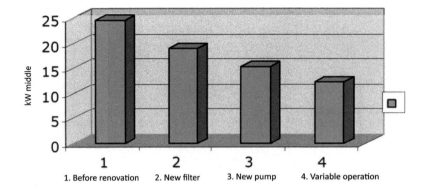

kW middle

1. Before renovation 2. New filter 3. New pump 4. Variable operation

3.47
Experience of improvement of pump systems in renovated swimming halls in Copenhagen has been transferred to the project. Improvements in kW electricity use are due to improved filter (2), new pump system (3) and variable operation (4), which can be compared to reference (1).

Efficient pumps

Filters with low losses

Optimised piping system

Regulation of water velocity

3.48
Electricity use has been reduced by efficient pumps, filters with low losses, optimised piping system and regulation of water velocity.

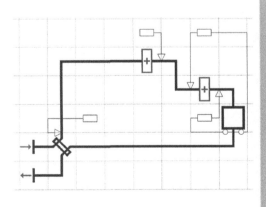

Ventilation with:

80-85% heat recovery

Recirculation in swimming hall

1,8 kJ/m³ SFP

Demand control

Optimised inlet efficiency

3.49
Example of a glazed facade heat loss calculation. Ventilation system with heat recovery and recirculation in swimming hall. Seasonal performance factor (SFP)/electricity use for fans is 1.8 kJ/m³. Demand control is installed and expected heat recovery efficiency is 80–85 per cent, as explained here.

PV installation: 19.11 kWp photovoltaic modules (128.4 m^2) are installed on the roof of an existing neighbour building on the same land register. The photovoltaics have been installed since 29 June 2012.

Damhusåen PV plant

Table 3.7
Damhusåen PV plant

Implementation	Finished January 2013
Plant size, kWp	777
Electricity production, kWh/year	700,000

3.50
PV panels mounted on the ground at 'Damhusåen'.

3.51
The PV panels were manufactured by Yingli.

The photovoltaic plant is positioned in Valby in connection to the waste water treatment plant. The photovoltaic plant is 777 kWp and is able to cover 8–10 per cent of the electricity consumption of the waste water treatment plant.

3.52
View over the 777 kWp photovoltaic plant.

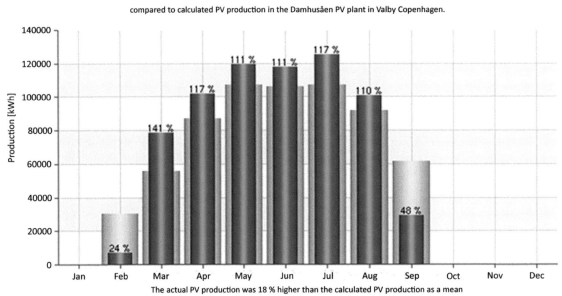

3.53
It is seen that a somewhat higher PV production than calculated has been obtained.

3.54
An opening event for the
large-scale PV plant in
Damhusåen waste water
treatment plant in Valby,
Copenhagen took place
in January 2013. A special
platform for people to view
the 777 kWp PV array (around
6,000 m²) was made for the
occasion.

3.55
On 30 January 2013 the
official opening event for the
Damhusåen PV plant was held
with speeches by the Mayor of
Environment in Copenhagen,
Ayfer Bakal, and the Director
of the 'Lynette Co-op', Mr
Torben Knudsen.

3.56
The builder was
'Lynettefællesskabet', here
shown with the EU Concerto
programme logo.

Lyshøjgård

Table 3.8
Lyshøjgård

Status	Finished early 2013
Plant size, kWp	70
Electricity production, kWh/year	62,000

A/B Lyshøjgård was realised in spring 2013, combining PV with new red-tile roofs
and extra roof insulation (200 mm).

3.58
Photovoltaic on the roof of the
Lyshøjgård shared ownership
housing in Valby.

3.59
Example of PV integration on new red-tile roofs in Lyshøjgård in Valby.

Table 3.9
Monitored electricity production for Inverter 5

Building [no.]	Inverter [no.]	Inclination [°]	Orientation [-]	Installed photovoltaics [kWp]	Installed photovoltaics [m²]
1	1	35	West	21.06	137.9
	2				
1	3	35	East	21.06	137.8
	4				
1	5	35	South	7.8	51
2	6	35	East	21.09	137.9
	7				

Valby Sports Hall

On the roof of the old Valby Sports Hall the Copenhagen PV Co-op has established 22 kWp PV modules. The PV electricity is sold to Copenhagen Energy/Dong Energy in a special Solar Stock Exchange feed-in tariff scheme and is recorded every month. All the photovoltaics are oriented south with a slope on 30°.

3.60
PV panels on top of Valby
Sports Hall.

3.61
The photovoltaic plant.

Peder Vejsig Pedersen

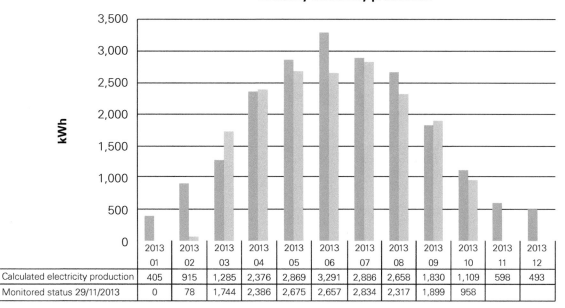

Monthly electricity production

	2013 01	2013 02	2013 03	2013 04	2013 05	2013 06	2013 07	2013 08	2013 09	2013 10	2013 11	2013 12
Calculated electricity production	405	915	1,285	2,376	2,869	3,291	2,886	2,658	1,830	1,109	598	493
Monitored status 29/11/2013	0	78	1,744	2,386	2,675	2,657	2,834	2,317	1,899	958		

3.62
Calculated and monitored electricity performance from February 2013 to October 2013.

Calculation of total energy savings and total economy for the Valby Project

Yearly energy balance for the EU Concerto area in Valby community

In the analysis we look at both the buildings that will receive EU support (67,143 m², as set in the EU–Green Solar Cities contract), and the greater area of low-energy buildings (80,553 m²) that was actually realised. This includes extra building areas of 13,410 m², which will not receive EU support, but are included in the Concerto demonstration programme.

PV output of 1,078 MWh comes from 1,198 kWp PV as the new total PV implementation size in the Concerto project in Valby. Multiplied by 2.5 we then have an energy frame value of 2,695 MWh. In addition there is solar thermal implementation of 387 MWh.

Therefore, in total we have 2,695 + 387 = 3,082 MWh = 68.2 per cent of the total building-related energy frame value of 4,520 MWh. If this only relates to the 67,143 m² building area, which is the official Concerto building area, the renewable energy contribution grows to 81 per cent.

An overview of energy frame values for different types of buildings is shown in Table 3.12.

Table 3.10

Renewable Energy Sources savings in Valby

		MWh
Common electricity use Concerto area	19,986 m² public/commercial building × 10	200
	22,230 m² retrofit housing × 6	134
	18,023 m² new-built housing × 6	108
	20,314 m² elderly people housing (new) × 15	305
	Total	747
District heating use Concerto area	19,986 m² public/commercial building × 23 kWh/m²	459
	22,230 m² retrofit housing × 41 kWh/m²	911
	18,023 m² new-built housing × 28 kWh/m²	505
	20,314 m² elderly people housing (new) × 43 kWh/m²	874
	Total	2,749
	Aimed at maximum heat losses in district heating, 10 per cent	275
District heating	Total	3,024

The new energy balance for the EU Concerto area in Valby is based on official energy frame value calculations from the Danish energy rules.

Table 3.11

Energy use in Concerto area in Valby as energy frame valve

			Building area m² (A)	Energy frame value kWh/m², year (B)	Energy use calculated as energy frame value MWh/year
1.	Water Cultural House		3,230	50	161
2.	Dr. Ingrids Plejehjem		11,671	63	735
3.	Langgadehus	family	5,823	33.3	193
		Elderly people home	8,643	101.9	881
4.	Karens Minde housing		4,300	66	283
5.	Hornemannsvænge housing renovation		22,230	55	1,222
7.	Teglholm School		1,260	50	63
8.	Ny Ellebjerg housing		7,900	36	284
9.	Lykkebo School		896	45	40
10.	Henkel II		13,500	45	608
11.	Ny Ellebjerg kindergarten		1,100	45	50
	Total excl. district heating losses		80,553		4,520 MWh

Table 3.12
Overview of energy frame values for different types of buildings

		Energy frame value MWh	Building area m²	Energy frame value in kWh/m², year	Of which electricity use: energy frame value and (consumption) kWh/m², year	District heating consumption kWh/m², year
1.	Public/commercial buildings	882	19,986	44	25/(10)	19
2.	Retrofit housing	1,222	22,230	55	15/(6)	40
3.	New-build housing	776	18,023	43	15/(6)	28
4.	Elderly people housing new-build	1,616	20,314	80	37/(15)	43

Table 3.13
Reference project

		MWh
Common electricity use	19,986 m² public/commercial building × 20 kWh/m², year	400
	22,230 m² retrofit housing × 12 kWh/m², year	267
	18.023 m² new-built housing × 12 kWh/m², year	216
	20.314 m² elderly people housing (new) × 25 kWh/m², year	508
	Total	1,391
District heating use	19,986 m² public/commercial building × 45 kWh/m², year	899
	22,230 m² retrofit housing × 160 kWh/m², year	3,557
	18,023 m² new-built housing × 56 kWh/m², year	1,009
	20,314 m² elderly people housing (new) × 80 kWh/m², year	1,625
	Total	7,090
District heating	Including 25 per cent net heat losses	8,863

For calculation of the energy frame value we have:

$$1,391 \times 2.5 + 8,863 = 12,341 \, \text{MWh}$$

The Concerto demonstration calculation as energy frame value is $747 \times 2.5 + 3,024 = 4,891$ MWh. This is only 39 per cent of the reference value, so 61 per cent has been saved only based on the building improvements.

Table 3.14
Savings in all in Valby communities based on new 2006/2010 energy rules in Denmark

Electricity in MWh	(1,391 – 747) MWh	644 MWh (54 per cent)
Electricity in euros	644 MWh × 227 €/MWh	€146,188
District heating incl. network losses in MWh	(8,863 – 3,024) MWh :	5,839 MWh (66 per cent)
District heating in euros	5,839 × 48 €/MWh	€280,272

Table 3.15
Total energy supply from renewables

	MWh
1,198 kWp PV × 900 kWh/kWp	1,078
860 m² solar collectors × 450 kW/m²	387

Table 3.16
Total renewables

Electricity	1,078 MWh (100 per cent of 747 MWh) and (1078 – 747) = 331 × 2.5 = 828 MWh as support for district heating
District heating/solar heating	387 + 828 = 1,215 MWh (40 per cent of 3,024 MWh)

Total investment for Valby community including design was €5,128,000; Table 3.17 shows the savings from solar energy.

Table 3.17
Savings from solar energy

		€
Savings from solar energy	PV electricity: 1,078 MWh × 227 €/MWh	244,706
	Solar heating: 387 MWh × 60 €/MWh	23,220
	Renewables in total	267,926

Peder Vejsig Pedersen

Table 3.18
Total economy (€)

Total energy saving value	267,926 + 280,272 + 146,188	694,386
Maintenance costs	0.5 per cent of investment	3,472
Real saving	694,386 − 3,472	690,914
Pay-back time	5,128/691 = 7.4 years excl. EU funding	

Taking the risk into account when using new technologies, this will lead to an acceptable overall economy for the project. It should be noted that part of the investment does not get funding from the EU. If this is taken into account the pay-back time is around ten years.

The overall results from the EU Concerto project in Valby, Copenhagen is that compared to the aim of realising a Concerto ecobuilding standard for 67,143 m² building area, it has been possible to realise this for 80,553 m² building area and with a 61 per cent saving compared to the normal reference standard in Denmark. Concerning the solar thermal collector area, the target was to reach 500 m², but 860 m² solar thermal collectors have actually been realised. Concerning PV systems, the target was to reach 441 kWp with the available EU funding; here 1,198 kWp have been realised due to the much-reduced PV costs in the project period.

If the energy frame value for the higher realised ecobuilding projects are compared to the solar energy input, then 68 per cent of the energy use is covered by solar energy. If you only use the original targeted ecobuilding area as a reference, then the solar energy contribution is 81 per cent. Looking at the pay-back time excluding EU funding for the original ecobuilding area, this becomes ten years, equal to seven years including the EU funding, which is reasonable when taking account of the risks involved.

Figure 3.63 shows overall heat saving analysed by the Steinbeiss Institute in Stuttgart for the realised EU Concerto project in Valby and Copenhagen.

3.63
Overall heat saving for the realised EU Concerto project in Valby and Copenhagen.

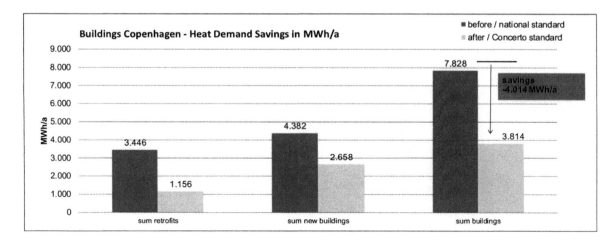

Table 3.19

Overview of EU Concerto Green Solar Cities individual demonstration projects in Valby in relation to overall energy savings and energy frame values shown as they are realised and compared to original expectations. The conclusion is that it has been possible to obtain clearly improved energy saving results as a general rule

EU Concerto, Green Solar Cities in Valby, Copenhagen Energy use in kWh/m², year		Existing	Heat + electricity Best sheet information				Danish national energy frame value			
		General national standard	EU Concerto 2008 annex	Demonstration project actual standard	Actual national standard	Improvement (%)	Actual building regulations	Demonstration Project Actual standard	Concerto specifications	Improvement (%)
1. New-build housing	(a) Karens Minde (BR06)	78	55	58/54/51(54)	78	31	101	56/50/48	77.5	49
	(b) Ny Ellebjerg (BR10)	78	55	38.9	58	33	66	35.6	77.5	46
2. New-build elderly housing	(a) Dr. Ingrids Hjem (BR08)	78	55	43	92.5	46	112	63	77.5	44
	(b) Langgadehus (BR06)	78	55	92.9	139.2	33	148.5	101.9	77.5	31
3. Rooftop dwellings	Langgadehus family housing (BR06)	78	38.5	32.5	83.5	61	101	33.5	58	67
4. Retrofit housing	Hornemannsvænge	178	54.5	46	155 (existing)	70	212 (existing)	52	67	75
5. Public building	(a) Teglholm/ Sydhavn school (BR10)	112	71	38.9	84	46	95.2	35.6	120.5	63
	(b) Lykkebo school/ sports building (BR08)	112	71	91.7	133	31	188	131	120.5	31
	(c) Henkel II (BR08)	112	71	26.7	112	76	95.3	66.6	120.5	30
6.	Water Cultural House Valby (BR08)	188	107	33.8	188	82	218	45.2	122	79

3.64
Site visit at Langgadehus by
Norwegian partners from the
EU Concerto project PIMES,
which also includes the city of
Szentendre near Budapest in
Hungary, which participated
as an associate city in the
Green Solar Cities project.

3.65
Promotion of optimised
ventilation solutions at Energy
and Environmental Theme
Days in Valby.

3.66
Site visit by Green Solar Cities partners to a rooftop common area in Copenhagen at a shared ownership apartment block.

Green Solar Cities: Salzburg

INGE STRAßL

The Concerto project Green Solar Cities has started and supported the renewal process in Lehen in Salzburg, where new homes and many other buildings have been erected. Through the consistent low-energy design, 40 per cent of energy per year could be saved. In addition, a 2,700 m² solar thermal system and 50 kWp photovoltaic system were built.

The district of Lehen is centrally located in Salzburg and has a good infrastructure. Most of the buildings were built between 1950 and 1970. In 2005, when the first plans for the Solar Cities project arose, there were many old houses in need of renovation, some brownfields and many shops, and social and public institutions that had moved away. Demographic trends and the many old residents of Lehen are also important. The adopted plan was to launch a comprehensive renewal process.

The old stadium was demolished and the site developed into a new centre with the construction of the municipal library, shops, cafes and 48 subsidised apartments. A 144 m² solar collector produces solar energy for hot water and heating.

In Esshaverstraße a low-energy house has been built in the area of a former mechanical workshop. Where earlier a Mercedes store was located, the project Parklife has been developed as the result of a design competition for young architects. Now, a nursing home with 90 rooms, 32 apartment for elderly people and 56 apartments for young families have been built. In the social centre there are many activities for the elderly and disabled, and it's not just for residents, but also people from the surrounding areas. Here, people can get a cheap lunch, play games in the afternoon, as well as other joint activities, and as a special feature there is a gym for people aged 70+, designed and equipped with special equipment. The buildings are all low-energy buildings with solar hot water and heating. On the ground floor there is a supermarket, shops and public green spaces.

For the old dwellings in Strubergasse a comprehensive redevelopment concept was created. The first phase of the thermal renovation was completed in winter 2012–2013.

Next to Strubergasse is the site of the former Salzburger Stadtwerke. Here, 287 apartments, the new city gallery, a dormitory and a kindergarten were built as part of the Concerto project. The apartments are in accordance with the low-energy housing standard, with controlled ventilation with heat recovery. The existing office building was renovated, and in the southern area offices, laboratories and seminar rooms will be built.

As the urban district heating system of Salzburg contains a very high proportion of industrial waste heat and biomass, it was important to find a system for the power supply that optimally complemented the urban district heating by solar energy. In the city of Lehen a solar thermal system was built with a 2,000 m^2 collector. The heat is collected in a central buffer with 200,000 litres. A solar heat pump optimises the system and gives an additional 15–20 per cent. The heat is distributed by means of a micro network to the apartments, offices and laboratories, as well as to the renovated houses in the area.

3.67
Demonstration project overview in Salzburg.

Stadtwerk Lehen

Design background for solar plant and micro network in Salzburg

The overall design was made by Steinbeis – Transferzentrum Energie-, Gebäude-
und Solartechnik from Stuttgart in 2011.

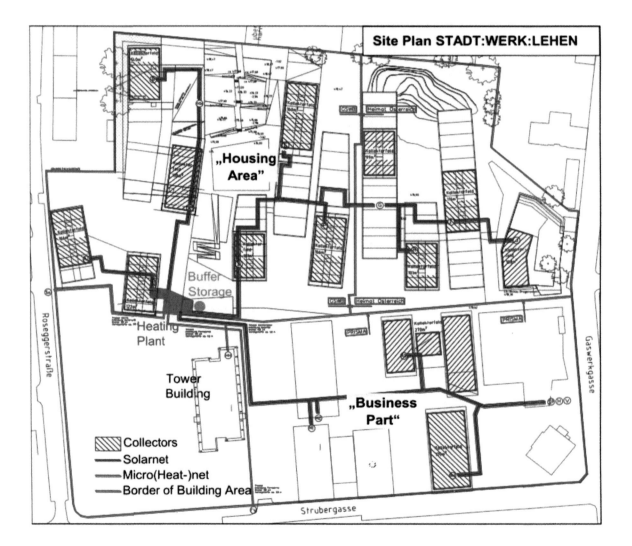

3.68
Overview of solar plant with buffer storage in Lehen.

Inge Straßl

3.69
Electrical heat pump: nominal heat output is 170 kW.

3.70
Buffer storage in front of the tower building.

3.71
Connection pipes and installation of the heat insulation and of the moisture barrier ('Klima Membrane') at the buffer storage.

Projekt-Nr.:	27311
Projekt-Name:	Concerto II, Stadt:Werk:Lehen

Cost of the Project:

Collectors (Fa. TISUN)	634 T€
Buffer Storage (Fa. BTD)	140 T€
Solarnet	166 T€
Micronet	169 T€
Solarstation in the Heating Plant	100 T€
Heat Pump with integration	75 T€
Measuring, Controlling and Electric	86 T€
District Heating Plant with Transfer Station	148 T€
Miscellaneous	35 T€
System Cost	**1,543 T€**
Design and Monitoring	133 T€
Total Project Cost	**1,666 T€**

3.72
Project cost break-down.
T€ = €1000.

Inge Straßl

Problems encountered

Intensive communication was necessary in order to realise a maximum solar area on the roofs. It was difficult to find solutions that were accepted by the designer (due to space for ventilation systems on the roof), architects (location and nice view) and building owners (clear distinction between solar part and the roof).

Due to the height of the buildings there is a significant wind load to be calculated. This caused some trouble regarding the boundary condition of the supporting structure.

During installation of the micro network some difficulties came up concerning crossing points of solar pipes and sewage pipes.

Several minor changes of the buffer storage had to be carried out during the final design phase.

Further comments on monitored results in Salzburg at Stadtwerk Lehen

The total investment costs of €1,543,000 can be divided into €900,000 for the complete solar thermal collector system and €215,000 for the heat pump and buffer storage, which mean a total cost of €1,105,000. There is an additional €86,000 for the energy management system and monitoring equipment, while the district heating and micro grid system cost are €352,000. Extra design costs were 8.6 per cent of the complete package.

With 2,047 m^2 of solar collectors the solar system investment, including storage, is €1,105,000/2,047, so €540 per square metre.

The operation results show that a low district heating return temperature from the buildings is obtained, at 35 °C, and that the 50 °C stratification between the top and bottom of the buffer storage is good, whereas there have been some problems when the heat pump was fed with higher entry temperatures (normally up to 20°C is best).

The overall heat use for the 45,164 m^2 building area was 38 per cent higher than calculated, primarily due to indoor temperatures being around 23 °C instead of 20 °C. Overall the target is to save 1,680 MWh in district heating, while 80 MWh of electricity is used for the heat pump.

The economic benefit from this is around €112,000; when compared to the previously mentioned costs this leads to an overall pay-back time of ten years without EU funding.

The targeted heat use was 60.8 kWh/m^2 per year of which 21.3 kWh/m^2 per year should be covered by solar thermal energy, leading to a total heat consumption of 39.5 kWh/m^2 per year.

Since only 1,551 m^2 of solar collectors were available in 2013 it was only possible to judge the final energy balance by summer 2014.

Stadtwerk Lehen residential part

3.73
High-rise housing with solar thermal collectors in the Lehen area in Salzburg.

Status	Dwellings	
Handed over 11|2012, monitoring	• 287 + Kindergarten and student dormitory • 36.117 m² gross floor area	
Thermal benchmarks	**Detailed U-values [W/(m²K)]**	
• THD 4-18 [kWh/(m² GFA)] • LEK 16.4-17.9	• Walls 0.13 • Roof 0.10-0.11 • Ceiling above ceiling 0.19-0.25 • Windows 0.69-0.89	
Housing technology		
Heating	Micro net + district heat + thermal solar	
Ventilation system	Controlled air ventilation with heat recovery in a central system	
Solar	1551m² + 200000 l buffer tank	
Part of the concerto project		
Building operations, solar plant 1551m²		

3.74
Illustration of status by November 2012 in the Lehen area.

Inge Straßl

General information

On the former premises of Salzburg's municipal utility, two non-profit housing developers (gswb and Heimat Österreich) built a total of 287 subsidised apartments from 2009 to 2011. They were handed over in November 2011. Furthermore, the area of 43,000 m² contains a student hall with 97 units and a kindergarten.

Energy supply

The largest thermal solar power plant in Salzburg – with 2,000 m² of solar collectors and a 200,000 litre buffer tank with an integrated solar heat pump – was installed in the district of Lehen. A grid of micronetworks distributes the generated solar energy to the new – and some of the old – buildings of the area. This system will cover about 30 per cent of the area's energy demand through solar energy.

Monitoring of the solar system and micro network

The Steinbeis Transferzentrum in Stuttgart was a partner in the Concerto project and responsible for the planning, simulation and monitoring of the solar system in Stadtwerk Lehen. A total of 310 different control points have been installed in the area, which deliver a range of information. Sensors control the temperature of the solar panels, feed and waste line of the heat pump and the district heating system as well as temperatures of the thermal distribution grid. Heat meters measure quantities of flow rates and capacity of solar plant, heat pump and amounts at heat transmission stations of the district heating network. The test phase from January to April 2012 was made to optimise the technical systems.

Heating energy demand calculated compared to real consumption

3.75
The heat demand is actually about 38 per cent higher than calculated. This is for several reasons: (1) room temperature for calculation was 20 °C, but the actual average is around 22 °C; (2) solar panels for calculation were to cover 2,000 m², but during the monitoring period this was 1,551 m².

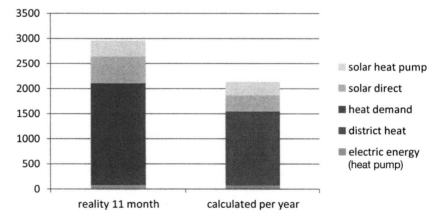

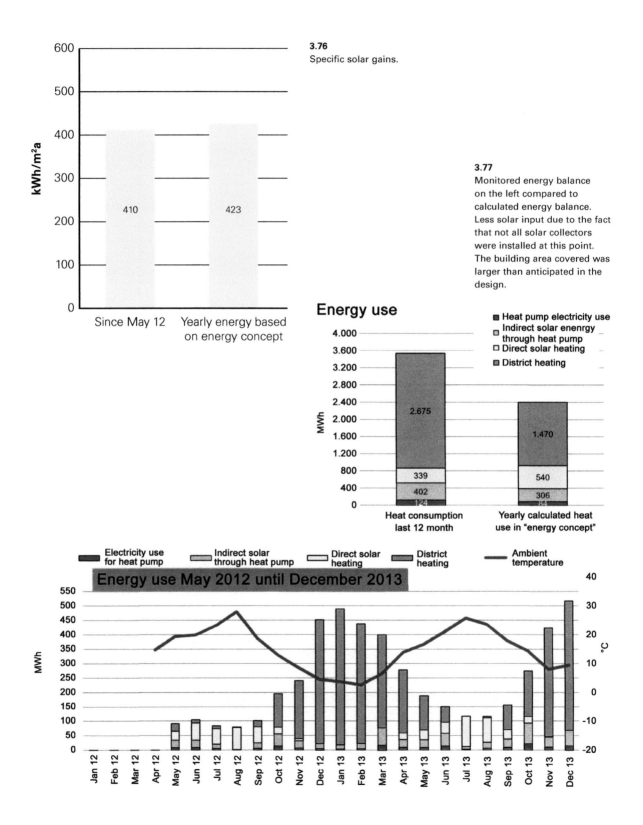

3.76
Specific solar gains.

3.77
Monitored energy balance on the left compared to calculated energy balance. Less solar input due to the fact that not all solar collectors were installed at this point. The building area covered was larger than anticipated in the design.

Energy use

Heat pump electricity use
Indirect solar enenrgy through heat pump
Direct solar heating
District heating

Electricity use for heat pump
Indirect solar through heat pump
Direct solar heating
District heating
Ambient temperature

Energy use May 2012 until December 2013

Solar system diagramme, Stadtwerk Lehen

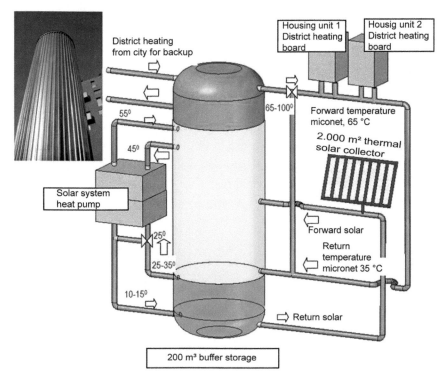

District heating from city for backup

Housing unit 1 District heating board

Housig unit 2 District heating board

Forward temperature miconet, 65 °C

2.000 m² thermal solar collector

55⁰

45⁰

65-100⁰

Solar system heat pump

Forward solar

Return temperature micronet 35 °C

25⁰

25-35⁰

10-15⁰

Return solar

200 m³ buffer storage

3.78
Scheme of the 2,000 m² solar system in combination with the micro network, 200 m³ buffer tank and heat pump, together with local district heating boards with domestic hot water heat-exchangers. Supplemental heat supply is from the city district heating network. Expected temperature levels are also shown.

The energy need in total is higher than calculated. Reasons for this are a change in the use of the ground floor area in one house and user behaviour – the average temperature in the apartments was calculated at 20 °C, but has actually been measured at nearly 23 °C.

The Steinbeis Transferzentrum continued the monitoring of the solar plant until summer 2014.

Visualisation

To inform and motivate the people living there or passing by, an LED visualisation was placed at the buffer tank. Here, the actual input of thermal solar energy and PV is shown (changing rather quickly) and provides the share of solar over the last 24 hours.

3.79
The 200 m³ buffer tank is a landmark in Lehen, Salzburg.

3.80
Diode indication showing results of the solar heating system.

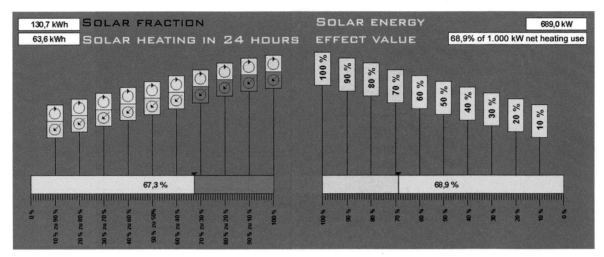

3.81
Indicators showing how much solar energy is provided in 24 hours to the left and in real time on the right; this is placed on the 200 m³ buffer tank in the Lehen district in Salzburg.

Conclusion

- The overall system (hydraulic) works very well.
- Specific solar gains are higher than expected.
- The return temperature of about 35 °C is very good.
- The principle of temperature spreading in the buffer tank via the heat pump works.
- The big buffer tank offers a very good temperature stratification (50 K).
- The heat pump has trouble with high entry temperatures (despite promises from the producer); this involves complex troubleshooting.
- Energy demand of the building is much higher than planned.
- Monitoring supports the optimisation of operations.

Energy monitoring

To monitor the energy consumption of the apartments and to test if (and what kind of) information can help to motivate people to save energy, a special research project has been launched in the course of a second research programme ('Haus der Zukunft') by the Federal Ministry of Transport, Innovation and Technology – intelligent E-monitoring.

The project partners implemented smart meters in 76 apartments. For research reasons the municipal utility Salzburg AG, gswb and SIR divided these into four groups, receiving different amounts and kinds of information on the energy consumption data. (a: reference group with no extra information; b: access to internet monitoring tool and receives monthly information on energy demands by mail; c: has access to an internet portal monitoring tool + additional monthly information on personal demands by mail; d: additional real-time data tool called 'Wattson' in the apartment).

Tenants are offered personal training on the use of their high-quality housing equipment, as well as a free energy consultation. SAG installed a 24-hour hotline in case of any technical or usage concerns.

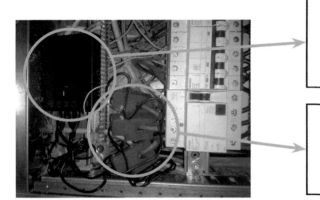

Transmitter
The transmitter is transforming the sensor signals and transfer this in a code by a radio signal to the Wattson display unit.

3 Sensors
The 3 sensors is continuously monitoring the electricity use in the apartment and transfer this data to the transmitter.

Wattson Display Unit
The Wattson Display unit receives data through a radio signal from the transmitter unit and shows the actual electricity use.
   Mean values are shown in purple and high values are shown in red.

3.82
The intelligent Wattson survey unit receives the data of actual electricity usage via radio and shows it with different colours (green – low; violet – average; red – high).

3.83
Introduction from GSWB of service personnel.

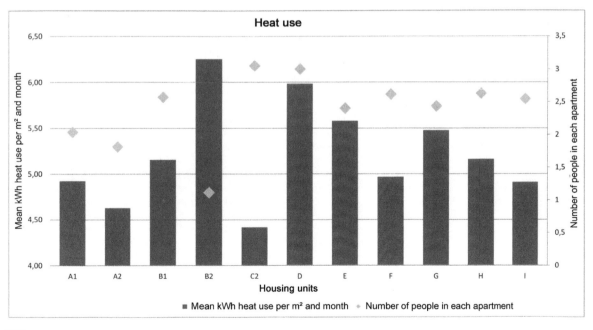

3.84
Red blocks: average kWh heat per square metre and month for different apartments. Green dot: Persons per apartment, so this does not explain the differences in heat use.

For this 'intelligent monitoring' no EU funding was used, but the results can help in the total overview of the project.

The monitoring shows that there is a wide difference within normal users, and no big difference between test apartments and 'normal' users. There is a 90 m² apartment needing 6,800 kWh for heating (= 75.5 kWh/m²) and there is a 77 m² apartment that used 1,730 kWh in the same time (= 18.0 kWh/m²).

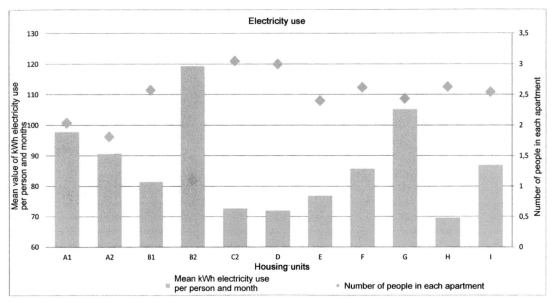

3.85
There is not much difference in electricity use between test apartments and the others.

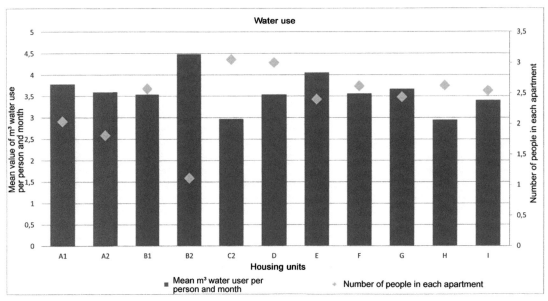

3.86
Water consumption is not so different.

The test apartments are in Blocks A1, A2 and B1, shown in Figure 3.84. This shows the wide difference in energy need for heating in the different houses. A2 is one block with 'test apartments', so there the energy use is a little lower than in most others. Block B2 is special because it contains only one-person apartments.

The system with the 'Wohnungsstationen' (district heating boards with a domestic hot water heat-exchanger and heat meter) has proved to work as a perfect system for large housing areas with very different users. Here, the heat is supplied to each apartment and the domestic hot water production is decentralised. The system has a perfect hydraulic regulation in all radiators and so user behaviour has no influence on the efficiency of the total system.

It can be concluded that the solar thermal system, after a 'running in' period, functions very well. The solar contribution is around 25 per cent due to more building area being connected to the system than first anticipated. At the same time the solar heating gain is higher than calculated, at 465 kWh/m^2 per year. The temperature difference between the upper and lower part of the buffer storage is 50°C.

In summer periods the solar system and heat pump provide most of the heat, but in the winter the district heating system covers 80 per cent of the heating demand. The performance of the heat pump at around 65°C in summer was as calculated with a performance factor (COP) of 3.4 to 4. In winter it reaches 4.2 to 4.6.

The buffer storage is heated to 90°C in summer. If the lower part of the storage is higher than 85°C this is cooled at night by use of the solar collector circuit.

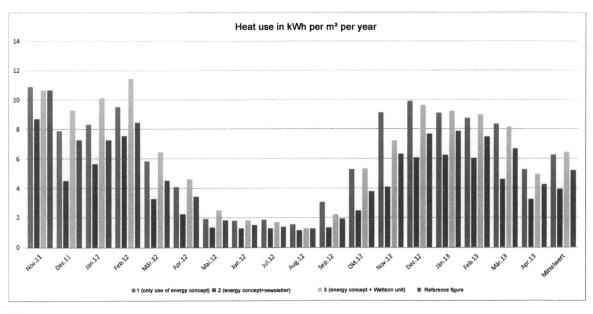

3.87
Heat consumption over one year. Blue indicates where just the energy concept is used; red is the energy concept with newsletters to tenants; green is the energy concept and Watson intelligent survey; and lilac is the reference figure.

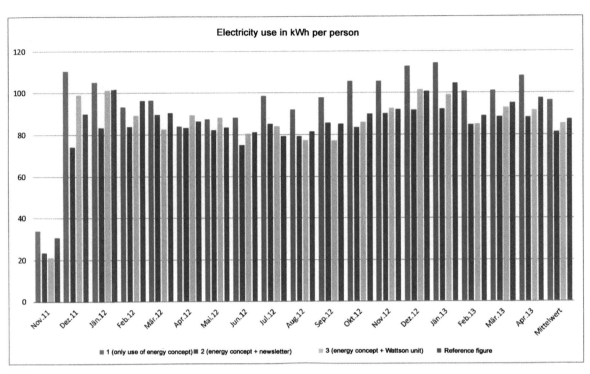

3.88

Electricity consumption over one year. Blue indicates where just the energy concept is used; red is the energy concept with newsletters to tenants; green is the energy concept and Watson intelligent survey; and lilac is the reference figure.

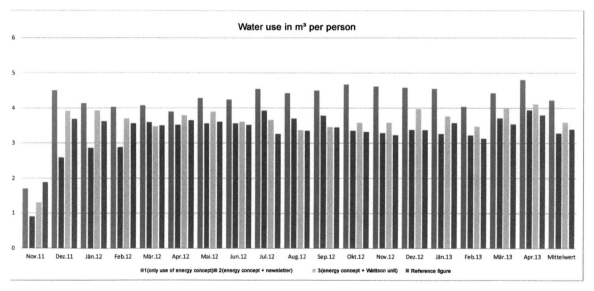

3.89

Water consumption in one year. Blue indicates where just the energy concept is used; red is the energy concept with newsletters to tenants; green is the energy concept and Watson intelligent survey; and lilac is the reference figure.

3.90
District heating board with 'small' domestic hot water heat-exchanger and heat meter.

Esshaverstrasse

3.91
Esshaver Strasse passive houses.

Status		Dwellings	
Finished 06 | 2008 • 1.072 m² gross floor area		• 12 apartments	
Thermal benchmarks		**Detailed U-values [W/(m²K)]**	

• THD calculated with GEQ = 1
  [kWh/(m²GFA)]
• THD calculated with PHPP = 34
  [kWh/(m²GFA)]
• Measured THD = 39
  [kWh/(m²GFA)]
• LEK 18
  (building law 38)
		• Walls 0.10–0.15
		• Roof 0.10
		• Ceiling above ceiling 0.10
		• Windows 0.71	

Housing technology			
Heating		District heat + buffer tank 4000l	
Ventilation system with heat recovery		Level of efficiency 87%	
Solar		38.55 m² (2.57 x 15)	
Part of the Concerto project			
Building work, thermal solar plant 38m², Detailed monitoring			

3.92
The Esshaver Strasse project was already finished in summer 2008.

General information

The four-storey building project Esshaverstraße 3 was commissioned by the non-profit building association 'Die Salzburg' and is located in Lehen, the central district of Salzburg. It contains 12 funded, rented and owner-occupied flats, which were completed in June 2008. The project was planned by Mayer und Seidl Architektur as a passive house

Energy supply

Heat supply is mainly covered by the district heating grid of the city of Salzburg. Additionally, 36m² of thermal solar heating was erected on the roof of the building, which supports the generation of domestic hot water as well as low-temperature heating through a storage tank. The use of solar energy annually creates about 11,900 kWh of heat energy.

Monitoring

The apartment building Esshaverstraße 3 was conceived following the passive house standard and calculated with software GEQ 2006. A further calculation with Passive House Planning Package (PHPP) software, as well as an analysis of monitoring data, showed that the value for limiting heat demand in a passive

Inge Straßl

house is exceeded. This particularly occurs on account of shading of window surfaces, orientation and shaded building elements and levels. A calculation done with GEQ insufficiently represented the influence of inner and solar gains. This discrepancy has been corrected in newer versions of the software. However, passive houses should always be calculated with an appropriate software – PHPP.

3.93
The official heat use calculation procedure (GEQ) in Salzburg showed a very low heat consumption (1.7 kWh/m² per year) compared to the passive house calculation tool. PHPP showed 34.4 kWh/m² per year, which actually fits quite well in the monitoring results (39.3 kWh/m² per year). The GEQ tool has been improved due to this experience.

Bestandsgebäude HWB$_{BGF}$ = 1,7 kWh/m²a **GEQ**

Calculated heat use with GEQ method

$\Big]$ **calculated**

Bestandsgebäude HWB$_{NGF}$ = 34,4 kWh/m²a **PHPP**

Calculated heat use with PHPP method

Bestandsgebäude HWB$_{NGF}$ = 39,3 kWh/m²a **MONITORING** — **measured**

Monitored heat use

Thermal heat consumption

Measured thermal heat consumption averages 38.2 kWh/m² per year for the whole building – about 11 per cent more than the 34.4 kWh/m² per year calculated by PHPP 2007. This overstepping can be explained: for example, during the heating period the average measured room temperatures were between 21 °C and 23 °C, whereas the typical value of 20 °C was used for the calculation. Another reason could be the fact that it is possible to heat the building outside of the heating season.

3.94
Indication of large differences between actual heat use (green) and calculated heat use (blue). Comparison of effective energy demand and consumption for each flat (kWh/m² per year). Green: effective energy consumption heating; blue: calculated effective energy demand (shaded).

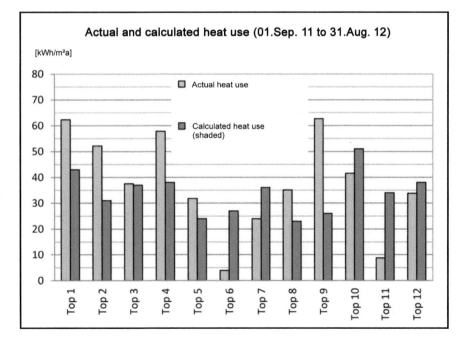

Actual and calculated heat use (01.Sep. 11 to 31.Aug. 12)

[kWh/m²a]

□ Actual heat use

■ Calculated heat use (shaded)

For energy-saving reasons it is recommended to implement zone valves in future projects, which enable the locking of heat supply during summer and transition times. Furthermore, the temperature efficiency of ventilation heat recovery plant does not achieve the required or manufacturer-guaranteed values, which have been used as the calculation basis. Measured values have to be verified and, if necessary, the system has to be rectified.

Hot water consumption

The average building hot water heating consumption is just over $15\,kWh/m^2$ per year and therefore slightly higher than assumed in the PHPP calculation. However, it was found that the surface-related heat requirement can further be used for the determination of the average hot water heat demand and also reflects the usual occupancy with sufficient accuracy.

Comfort parameters in the apartments

The temperature is comfortable in all apartments in winter and was evaluated as very high. In summer, inside temperatures of top-floor flats are sometimes a little high. In winter the room humidity is at a minimum of 25–30 per cent relative humidity over short periods, and therefore complies with inside air humidity in exclusively window-ventilated buildings. Therefore, no reduction in comfort or additional drying of the air can be determined due to the ventilation system.

When it comes to residents' ventilation behaviour, measurement data show the combination of air-conditioning systems and window ventilation, both in summer and winter. There is no proof of a correlation between window ventilation and an increased thermal heat demand. The possibility to operate the ventilation plant as needed with different intensities is perceived as necessary by the tenants; the three-step multi-level scheme has proved to be a success.

System engineering

By checking measured data and tracking of faults, weak points of the control system and insulation of the service plant can be discovered and rectified. It is recommended to use fortified insulated ducts in future projects.

The solar plant covers about 22 per cent of the total end energy consumption and provides a specific solar entry of $422\,kWh/m$ per year, which is significantly beyond the requested $350\,kWh/m^2$ per year of the housing fund of Salzburg. The plant operates reliably and delivers a precious share of the overall energy demand of the building Esshaverstraße 3.

Refurbishment Strubergassensiedlung

3.95
The renovated Strubergassensiedlung.

3.96
189 apartments were renovated with new facades and windows and connected to the micro grid.

Status	Dwellings
Finished	• 189 apartments • 16.660 m² gross floor area
Thermal benchmarks	**Detailed U-values [W/(m²K)]**
• THD 24 [kWh/(m² GFA)] • LEK 30	• Walls 0.23 • Roof 0.15 • Ceiling above ceiling 0.35 • Windows 0.9
Part of the Concerto project	
Retrofit and connection to micro grid	

Thermographic exposures

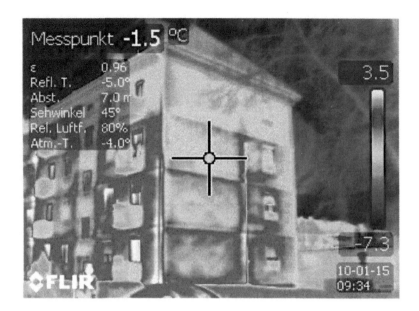

3.97
Very high surface temperatures in thermo photos at all connections of floors, walls, etc. before renovation, shown by the large differences in colour.

3.98
'After renovation', with much more uniform thermo photo colours.

3.99
Night view.

The dwellings are partially at different temperatures. All details in the renovated facade construction were made to a high standard, with the result that no thermal bridges are visible.

The heat saving in Salzburg is shown from analysis by Steinbeiss Institute in Stuttgart (Figure 3.100).

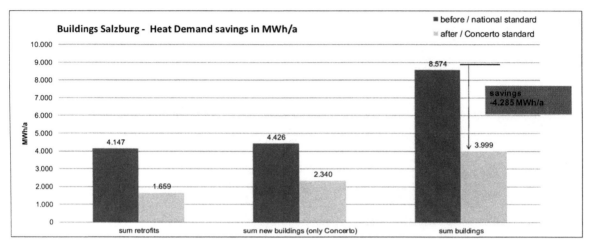

3.100
Heat demand savings for Salzburg buildings (MWh/a).

Best-practice examples

Some best-practice examples from Salzburg were presented at the Green Solar Cities final conference in April 2013 in Salzburg.

3.101
Passive house project in Salzburg near the railway, visited at two Green Solar Cities partner meetings and conferences in Salzburg.

3.102
The Stiegelgründe low-energy housing was also visited by the Green Solar Cities partners. It is consigned with large solar thermal collectors and use of biomass in the form of wood pellets. The wood pellet storage is integrated with the housing blocks and can last for two weeks in winter periods.

3.103
Best-practice smart grid housing area example from Salzburg presented at the site visit in April 2013.

3.104
Visit at a low-energy housing project with only demand-controlled exhaust ventilation, here seen in the kitchen. Some of the tenants were not happy about the indoor air quality of this solution.

3.105
Inge Straßl from SIR giving an introduction to the Green Solar Cities partners before the bike tour to best-practice projects.

3.106
Architect and politicians being interviewed at the final Green Solar Cities conference in Salzburg.

3.107
Green Solar Cities conference, April 2013, in Salzburg.

3.108
Large solar thermal collectors for district heating placed on the roofs of the housing blocks in Lehen, Salzburg.

3.109
The buffer storage in Lehen area in Salzburg has a sculptural function in the area.

3.110
Green Solar Cites partner meeting in Salzburg, spring 2013.

Chapter 4

Energy-Efficient Buildings

PEDER VEJSIG PEDERSEN AND KATRINE VEJSIG PEDERSEN

Passive house development in Sweden, Austria and Denmark

The following information is primarily from a report of the EU Concerto Green Solar Cities project, 'Passive house standard in Sweden, Denmark and Austria', by Professor Maria Wall from Lund University, Division of Energy and Building design.

There is an interesting introduction on how the idea of the passive house was first developed by the Swedish professor Bo Adamson of Lund University in the 1980s, and later moved further into practice by Wolfgang Feist of Germany, who also developed the Passive House Institute.

In Germany more than 10,000 passive houses have been established.

Passive house development in Sweden

Sweden has a long tradition of development of low-energy buildings, but during the 1990s the development almost stopped due to a period of poor economy and very limited production of new buildings. When countries like Germany, Austria and Switzerland developed passive houses and low-energy buildings, Sweden was therefore falling behind. However, an important step was taken with the development of the first passive houses in Sweden, which were the first in the Nordic countries.

The Swedish passive house concept

A passive house has a highly insulated and airtight building envelope combined with a mechanical ventilation system with efficient heat recovery. This means the space heating at peak load can be so low that the supply air is enough to distribute the heat into the building, using comfortable airflow rates. The energy source to preheat the air could be electricity, preferably coming from a renewable energy source like wind power or solar PV, or it could be heating, where district heating, biomass or solar heating could be good options.

In the present passive house criteria in Sweden, the total heat losses (fabric + ventilation/infiltration losses) for residential buildings should be between 15 and 17 W/m^2, depending on climate zone. For buildings smaller than 400 m^2 the maximum losses should be 17–19 W/m^2. In Germany the space heating peak load is at maximum 10 W/m^2 (after reduction of solar and internal gains), and this value can thus not be directly compared with the Swedish version. The Swedish definitions for passive houses can be found at www.nollhus.se/index.php/kriterier.

The development in Sweden started with the Lindås project

In 1997 a pre-study was carried out to find out whether it would be possible to build a passive house in a Swedish climate. The study was made for the preliminary design of terraced houses in Lindås outside Gothenburg. The climate in Sweden is colder and less sunny than the Central European climate. This makes it harder to achieve the passive house standard. The pre-study showed that it would be possible to build passive houses in the Gothenburg area. A project team was organised and research funding secured so that the work could continue to design, construct, monitor and evaluate the building performance.

4.1
Passive houses in Lindås, Gothenburg, built 2001. Photo: Maria Wall.

Peder Vejsig Pedersen and Katrine Vejsig Pedersen

The passive houses were completed outside Gothenburg in Sweden in 2001. Twenty terraced houses were built according to the passive house standard. Monitoring and evaluation of the completed buildings showed that the average delivered energy requirement was 68 kWh/m^2 per year, including DHW and space heating, electricity for fans and pumps and household electricity. This is approximately 40 per cent of the energy demand for average single-family houses in Sweden. The solar collectors contributed 8.9 kWh/m^2 per year to the DHW heating, which means the total energy requirements for the houses were approximately 77 kWh/m^2 per year.

The Swedish building code set a maximum energy requirement for DHW heating, space-heating and electricity for mechanical systems at 110 kWh/m^2 per year in the southern region, but the houses in Lindås only need approximately one-third of this (36 kWh/m^2) (see Figure 4.2). Thus, it is possible to build houses with a much lower energy demand than required by this building code (five years newer than the passive houses!). The Swedish building code has become somewhat stricter after this version.

The lessons learnt from this first project spread to new developers and led to new passive houses being planned and built. There was extensive research,

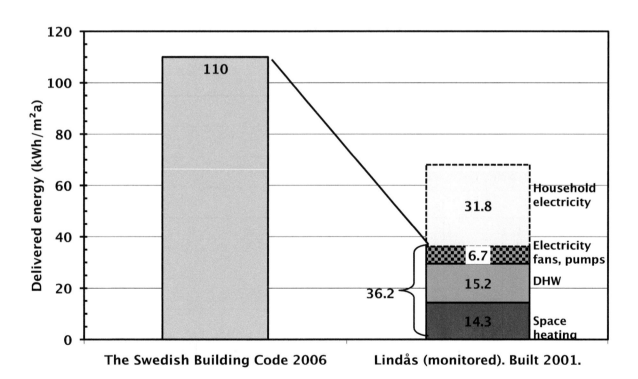

4.2

Monitored energy use for the terraced houses in Lindås (right). Delivered energy for DHW and space heating, and electricity for fans and pumps (36.2 kWh/m^2 per year) is compared to the requirements in the Swedish building code (left) for the southern climate zone (110 kWh/m^2 per year).

including documentation of the construction of passive houses, such as the first single-family house in Sweden, new apartment buildings and also a renovation project with apartment buildings from 1970 built in Alingsås (Brogården), in which the goal is to reduce the energy use by approximately 50 per cent compared to today.

Passive house development in Austria

Austria was quick to import the passive house concept from Germany. The passive house market in Austria has rapidly increased since 1996 (www.igpassivhaus.at) (Figure 4.3). At the end of 2006, 4 per cent of all new buildings were passive houses, and estimated share of passive houses in 2010 is 24 per cent; the estimate for 2020 is 62 per cent. The majority of Austrian passive houses are single-family houses. Vorarlberg has the highest density of passive houses in Austria. Also, renovation to the passive house standard is growing rapidly. Government housing subsidy programmes have a direct impact on the number of residential buildings built as passive houses – for example, in Vorarlberg housing subsidies are granted only if the buildings meet the passive house standard.

The development of low-energy buildings and renewable energy supply are supported by the Austrian 'Programme on Technologies for Sustainable Development', which funds future-oriented innovations and developments. This five-year research and technology programme was developed by the Austrian

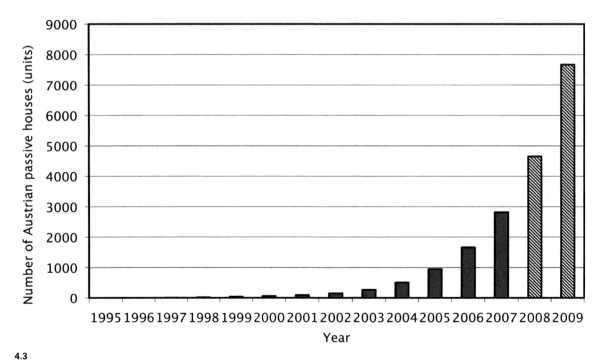

4.3
Documented passive houses in Austria since 1995 and predicted for 2008–2009 (data from www.igpassivhaus.at).

Peder Vejsig Pedersen and Katrine Vejsig Pedersen

Federal Ministry of Transport, Innovation and Technology (BMVIT). It initiates and supports research and development projects and the implementation of exemplary pilot projects. Three sub-programmes exist: 'Building of Tomorrow', 'Energy Systems of Tomorrow' and 'Factory of Tomorrow'. More information can be found at www.hausderzukunft.at/english.htm.

The city of Salzburg

The city of Salzburg is one of the cities included in the Green Solar Cities EU Concerto II project. In Austria each federal state is responsible for its own subsidised housing. Since 1993 Salzburg has used an incentive scheme for energy-saving measures in buildings and the use of renewable energy. Based on the assignment of energy points, the system increases the subsidy funding. The reward system reaches a very high percentage of newly constructed residential buildings. The achieved points depend on the energy performance and ecological quality of the building. Buildings are divided into ten classes and include passive buildings (see Deliverable 2.2: Reward system for ecologic housing in Salzburg. A point-based system for the funding of energy-efficient eco-buildings and the use of biomass and solar energy. Version 1, 23 May 2008). This reward system has proved to be successful in the implementation of energy-efficient and sustainable buildings.

As part of the Green Solar Cities project, the passive house project in Esshaverstraße 3 in Salzburg was the first completed residential building. The

4.4
The passive house apartment building in Esshaverstraße 3 in Salzburg.

Concerto site is located in the Lehen district of Salzburg. The passive house building includes 12 rental apartments and was completed in June 2008. It promotes the improvement of energy efficiency and the use of renewable energy. Salzburg supports such targets and declared the project to be a residential role model of the state. The apartment building meets the passive house standard class 10 according to the subsidised housing decree in Salzburg. Furthermore, it is one of the first social housing buildings in Salzburg constructed as a passive house. The building was monitored to evaluate the performance (see pp. 128–130).

Passive house development in Denmark

Rønnebækhave II

Denmark started the development of passive houses later than Sweden. The first Danish passive house building project, Rønnebækhave II in Næstved, south of Copenhagen, was finished in February 2006. It comprises a small housing block with eight apartments realised according to the German passive house principles.

To be able to live up to the German passive house criteria with a space heating demand not exceeding 15 kWh/m^2 per year it was necessary to purchase low-energy windows from Germany with a documented U-value of 0.8 W/m^{2}K and an individual only 26 cm thick Ecovent heat recovery unit to enable a 'dry' heat recovery efficiency of 85 per cent, together with very low electricity use. For two of the apartments ventilation air is preheated in the ground.

The apartments utilise a 28 m^2 shared solar DHW system and individual 240 litre DHW tanks. A ground-coupled shared heat pump is used in combination with air heating and floor heating. This is matched by PV electricity from a 5 kWp PV system in the roof to achieve an annual CO_2-neutral heating design.

With a calculated energy use for heating and DHW of 23 kWh/m^2 per year, the electricity need for the heat pump with a COP of 3.5 will be only $23/3.5 = 6.5$ kWh/m^2. This is the same as the yearly PV production of the PV modules, which is 4,250 kWh or 6.6 kWh/m^2 per year.

The experiences with the passive housing project in Næstved have been very good. In particular, the balanced HRV system is liked by the tenants and actually two families claim they have avoided asthmatic symptoms since moving into the apartments.

Due to the limited knowledge concerning passive houses in Denmark at that time, and the general policy concerning energy, it was very difficult to get the project financed and it was also difficult to get funding for monitoring and follow-up. But monitoring from spring of 2007 shows that it has been possible to obtain results very close to what was predicted, with an energy use for heating and DHW near 20–25 kWh/m^2 per year. The actual COP value of the heat pump was measured at 3.2–3.4, which is reasonable compared to expectations.

Peder Vejsig Pedersen and Katrine Vejsig Pedersen

Further development of passive houses in Denmark

Another passive house was finished in spring 2008; a single-family house, 'Langenkamp' in Ebeltoft (www.langenkamp.dk). This is the first Passive House Institute certified passive house in Denmark. New passive house projects are developing and more information can be found in Danish on the websites: http://vbn.aau.dk/en/publications/living-in-some-of-the-first-danish-passive-houses(4055f992-3a92-49ef-91b1-b589a304fdae).html, www.altompassivhuse.dk and www.passivhus.dk.

Monitoring results of Langenkamp passive house in Ebeltoft in Denmark have been provided by Kuben Management Energy Key monitoring system (funded by the Solar City Horsens EFP project). For a monitoring period from May 2010 to March 2011 the following results can be presented:

1. The small Austrian heat pump used 2,420 kWh in electricity and delivered 5,740 kWh in heat equal to a yearly COP of 2.4. The electricity use for ventilation was between 42 W and 83 W for the German HRV air heating system, with a mean value of 56 W (the airflow in winter is higher than in summer).
2. The room heating demand for the ten months is quite low at 3,247 kWh (20 kWh/m² per year). But the domestic hot water use is high at 4,879 kWh, of which the solar heating system covers 49 per cent.
3. Overall the house seems to have functioned reasonably close to the intentions. With an improved heat pump and ventilation systems reaching COP of 3.5 and a maximum 35 W in mean electricity use for ventilation, the electricity use for heating and ventilation could be reduced by 33 per cent, leading to a yearly energy frame value of 36 kWh/m² per year, which in this case is a little better than low-energy class 2015.

4.6
Villa Langenkamp in Ebeltoft in Denmark.

4.7
German-style passive houses
in Hannover Kreutzberg before
EXPO 2000.

Peder Vejsig Pedersen and Katrine Vejsig Pedersen

4.8
Rasch and Partner was one of the first passive house developers in Germany, aiming at a low price and a low energy use.

Zurich

4.9
Passive house renovated housing block in Zurich.

4.10
Facade of passive house renovated housing block in Zurich.

4.11
Integrated PV example in Zurich, where 'solar electricity' has been sold to environmentally concerned consumers a little overpriced for a number of years (Solar Stock Exchange). The combined use of passive house technology and PV electricity is, however, not seen so often.

4.12
Three-dimensional biological
garden in Zurich; also a Green
Solar Cities Approach.

4.13
Low-energy housing area in
Zurich, built according to the
'Miniergie' standard.

Practical experience with solar low-energy building from the 1980s onwards

Solar thermal collectors

4.14
Solar thermal collectors investigated with respect to reliability and durability at the Thermal Insulation Laboratory in Denmark in the 1980s.

4.15
The director of Arcon Solar Heating, Sven Andersen, mounting large solar collector areas on a test roof at DTU.

Peder Vejsig Pedersen and Katrine Vejsig Pedersen

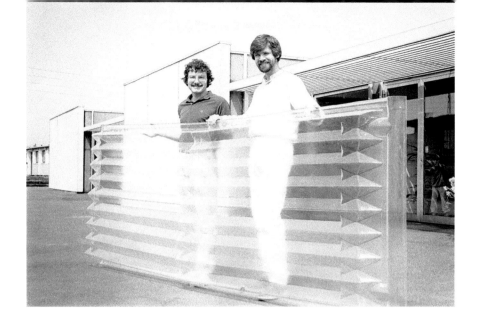

4.16
Two of the partners in
Cenergia, Peder Vejsig
Pedersen (to the right) and
Ove Mørck, presenting an
innovative solar collector
glazing system in 1982.

*Practical experiences with innovative compact thermosyphon solar
water heater*

4.17
In 1980 an experiment was
established at the Thermal
Insulation Laboratory with a
solar thermosyphon system
for domestic hot water, where
the domestic hot water tank
was placed at the same
level as the solar collector.
The result was a slow but
efficient operation with a good
temperature stratification. To
avoid backwards operation at
night a special valve with no
pressure loss was imported
from the USA and used to
secure this.

4.18
By using inventor-fund money, the compact solar water heater was further developed including a unique solar collector acrylic glazing system, which absorbed the solar input more efficiently. The concept was further developed for 55 dwellings in Snekkersten, north of Copenhagen.

4.19
Full-scale tests of an innovative thermosyphon solar water heating system. Installation of the efficient compact thermosyphon solar water heater system in Måløv near Copenhagen.

4.20
Mounting of building-integrated solar collectors in Måløv.

The idea of placing the solar DHW tank in the attic proved not so reliable as three leakage events were experienced: one frost leakage just after installation; one frost leakage from the water/air heat-exchanger used for room heating; and a leakage after 20 years when the DHW tank corroded. Now a DHW tank has been installed outside the house using a normal pump. The roof orientation was 20° from the east towards the south and even though the solar collector area was twice as large as necessary to be able to provide room heating in winter, it gave very little room heating due to the orientation and a lot of overheating in summer due to the high efficiency and size. Tank temperatures up to 110°C were normal. Today only half the solar collector area is used, but it still works well after 32 years of operation.

Fifty-five solar low-energy houses in Snekkersten north of Copenhagen (Figure 4.21) were energy-designed by Cenergia and established with an innovative building-integrated thermosyphon solar water heater integrated into the south-facing wall. Theoretical calculations showed a low heat consumption due to use of heat recovery ventilation in the houses. Based on this it was chosen to use electric heating. Due to the large variation in heat use, the outcome of this choice was different for different house owners. But for some, it was a costly solution.

4.21
Smakkebo low-energy houses in Snekkersten.

4.22
Innovative thermosyphon
solar water heater integrated
in the wall.

Peder Vejsig Pedersen and Katrine Vejsig Pedersen

Figure 4.22 offers an illustration of the wall-integrated thin domestic hot water tank and the undersized expansion tank in an energy-efficient compact solar water heater. Due to the obtained high temperatures of the solar heating systems the solar fluid boiled several times and in the end a more normal solar heating solution for the houses was introduced. Lesson: always make a full-scale test of innovative solutions in one housing unit, not in 55.

Solar thermal collector experiences

The Havrevangen low-energy housing project was built in the mid-1990s, based on a Nordic low-energy building competition. The project included a large air solar collector working in correspondence with floor heating. It was organised with KAB Housing Association in Hillerød, with good operational results.

4.23
Air solar collectors in Hillerød.

4.24
EU-supported demonstration project at Østerhøj in Måløv near Copenhagen, using a roof-integrated solar heating system.

4.25
Innovative solar collector roofing system from Sweden at Egebjergvang in Ballerup.

Peder Vejsig Pedersen and Katrine Vejsig Pedersen

4.26

The Nordic solar hot water demonstration system at Egebjergvang in Ballerup was established with the KAB housing association in 1984. Here, an innovative glazing system from Sweden was introduced with success.

4.27

The Swedish solar collector glazing system was demonstrated several places in Denmark. Here, on a housing block in Roskilde.

4.28
Example of test design
for building-integrated
ventilated solar walls. Here,
Grønnegården in Nykøbing
Sjælland.

Large solar thermal plant at Tubberupvænge II in Herlev near Copenhagen

The solar low-energy demonstration project Tubberupvænge II in Herlev included glazed communal areas for the tenants and integrated solar collectors built according to newly established BPS design rules, where thermal solar collectors were mounted on top of an asphalt-layer roof and surrounded by normal tiles.

Peder Vejsig Pedersen and Katrine Vejsig Pedersen

4.29
The Tubberupvænge II EU-supported solar low-energy housing project with 100 housing units was finished in 1990.

4.30
Building-integrated solar thermal collectors in Tubberupvænge in Herlev.

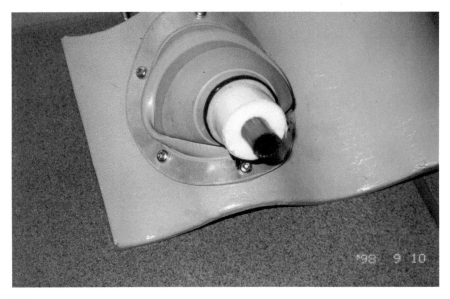

4.31
BPS – a bespoke solution for pipes penetrating the roof construction.

4.32
At Tubberupvænge II there was 1,000 m² of highly efficient Arcon Solar Heating solar collectors developed for district heating, and besides this 400 m² of local solar collectors for domestic hot water. As a new solution a 3,000 m² seasonal storage was introduced.

Peder Vejsig Pedersen and Katrine Vejsig Pedersen

4.33
Cheap pit storage based on
RTD work at the Thermal
Insulation Laboratory
in Denmark. The design
engineers were afraid of the
reliability when the clay wall
reached temperatures up to
90–95 °C in summer.

4.34
Seasonal storage with
concrete walls during
construction.

Even with a 3,000 m² seasonal storage area, the surface/volume ratio is quite high, so it was necessary to use good insulation and combined use of a heat pump to cool the storage to 5–10 °C in winter to avoid high heat losses. At seasonal storage sizes from 50,000–100,000 m² the heat losses will be much reduced.

4.35
Finalised seasonal storage
with a rubber liner. This was
the first seasonal storage
project realised in Denmark.

A traditional storage design with a concrete construction and a rubber liner from the US company Carlisle resulted in leakage of water, which was identified after 1.5 years of operation. Two years later the liner was changed to a stainless steel liner based on Swedish technology.

4.36
Solar collectors integrated into
the roof in a reliable way.

4.37
In Tubberupvænge II prefabricated housing units were used. Unfortunately, the demand for good airtightness was not introduced efficiently for the prefabricated housing units.

4.38
Mounting of the 12.5 m² large solar collectors along the Klausdalsbro road in Herlev, near Copenhagen.

4.39
Since 2005 large solar collectors for district heating have increased their market in Denmark. Plants with up to 50,000 m² of solar collectors have been established for local district heating systems, with a typical 20 per cent solar contribution.

Skotteparken solar low-energy housing in Egebjerggård, Ballerup

An EU-funded project was realised in 1992 in the neighbouring town of Ballerup at Egebjerg, again with the KAB Housing Association, and with the same architect and engineers as in Tubberupvænge II (Hanne Marcussen and Jens Peter Storgaard, together with Domnia engineers). In addition, the same contractor was involved. The overall quality was much better, but airtightness was still not good enough and an idea of insulating air ducts in the attic with loose insulation granulate did not work well since leaked indoor air could condense on the cold duct parts and moisture gather on top of the attic membrane.

At the Skotteparken development in Egebjerggård in Ballerup there was a high focus on ensuring good airtightness in cooperation with the contractor. But this was ruined by the electrician when he mounted all electric cables in the air gap between the plastic membrane and the tile wall. Here, plenty of holes for the electric cables ensured a completely uncontrolled air leakage. This was in 1992, almost at the same time as the first passive house was built in Germany showing how important it was to avoid air leakages.

4.40
Prefabricated housing elements were assembled on site. They were airtight until the electrician drilled holes from the outside for electrical cabling.

Peder Vejsig Pedersen and Katrine Vejsig Pedersen

Heat Demand Halved in Danish Buildings

Skotteparken is a non profit housing project that utilises a high degree of solar heating.

Electricity Consumption by Utilisation 1972 - 1994

Source: Energy Statistics 1994, The Danish Energy Agency.

Total electricity consumption has doubled since 1972. In recent years, however, the increase has been relatively low.

Since 1972 energy consumption for heating purposes in Danish buildings has dropped by more than 50% per m². At the same time many buildings have converted from oil to more environmentally benign natural gas and district heating. The result today is that almost two-thirds of all buildings in Denmark are connected to natural gas or district heating networks.

The greatest drop in energy consumption was achieved in the early 1980s. This was partly the result of state subsidy of thermal insulation and other energy-saving initiatives. It is of growing importance for space heating saving that the authorities have made increasingly stringent demands on energy conservation in new buildings. The latest Building Regulations (1995) require that space heating consumption should be reduced by a further 25%.

Continuous efforts are being made to reduce energy consumption in existing buildings with the help of consultants who are specially trained in energy management, and who are able to advise owners and staff on energy efficiency, energy-saving investments, and methods of energy management.

Ecological Building with 50% Lower Energy Consumption

The occupants of a building benefit when a developer or building society invests more money in energy conservation at the planning and construction stage, because depreciation of this increased expenditure is balanced by lower running costs. This is the experience gained from "Skotteparken", a building project in Ballerup near Copenhagen. The developer here was Ballerup Ejendomsselskab together with KAB, Building and Housing Management.

The 100 dwellings at "Skotteparken", where the first tenants moved in during 1993, entailed a cost-

effective investment about 8% higher than usual, because of the energy saving measures that were initiated.

Measurements taken during the first two years show that energy consumption has been cut by half, and water consumption reduced by a third, compared to other buildings. Cenergia Energy Consultants were responsible for the combined utilisation of low energy design, solar heating, and local combined heat and power generation (CHP) with low temperature pulse operation of the district heating network, which is demonstrated at "Skotteparken".

The project has received the international housing prize: "The World Habitat Award", and experience gained from "Skotteparken" will be applied in European network cooperation among housing companies. "The European Housing Ecology Network" is to erect 11 solar-heated, low-energy buildings in 7 EU Member States with the support of the THERMIE programme.

Skotteparken was designed by architects Hanne Marcussen and Jens Peter Storgaard together with the consultant engineering firm of Dominia A/S.

4.41
Example of promotion of the Skotteparken project, which won the World Habitat Award 1994.

4.42
Metallic roofs with solar collectors in Skotteparken.

4.43
View of one of seven technical
rooms in Skotteparken.

Peder Vejsig Pedersen and Katrine Vejsig Pedersen

With the help of Energy Management Control, a smart operation of the district heating supply was made in Skotteparken, so all tanks were filled with supplemental heat; if one tank called for heat, all the pipes in the ground were filled up with colder temperature return water to reduce distribution heat losses.

The solar low-energy Skotteparken development with 100 housing units at Egebjerggård in Ballerup was actually a great success and proved how large solar collectors could be coupled with district heating (here with 7 m^2 solar collectors per dwelling). And even though the airtightness could have been better, the energy use for heating and DHW was reduced from 180 kWh/m^2 per year to 70 kWh/m^2 per year according to monitoring results. In 1994 the project won the International Habitat Award and it formed the background to establish the European Housing Ecology Network (EHEN).

All solar thermal collectors have an orientation between southeast and southwest. This is important in order to allow solar heating not only for domestic hot water but also for heating.

Egebjerggård
– total energy design
for 100 dwellings

15 minutes
© ETHELBERG FILM

4.44
View of Skotteparken from the air

4.45
The solar heating storage tanks were built into the technical rooms during construction.

4.46
To avoid sending water to the sewer, it was instead led to a small lake in Skotteparken.

Peder Vejsig Pedersen and Katrine Vejsig Pedersen

Also at Egebjerggård a solar low-energy housing project with efficient windows was introduced in combination with combined heat recovery ventilation and an air-based heating system. Unfortunately, no radiators at all were used and the air heating proved difficult to operate and control in an efficient way for the users.

4.47
Agernskrænten solar low-energy housing designed by Henning Larsen Architects.

The Agernskrænten development with 50 housing units in Egebjerggård in Ballerup municipality was realised with EU funding from the EHEN project. It utilised solar thermal collectors on the flat roofs, use of a new type of Velfac three-layer glass windows and a combined heat recovery ventilation/air heating system, which actually proved how difficult it is to make this type of combined system in a successful way.

4.48
Smart air inlet solution at Agerskrænten low-energy houses at Egebjerggård in Ballerup.

Architectural competition on PV-assisted ventilation

Lundebjerggård, Skovlunde in Denmark

In 1997 a large architectural competition was held on energy renovation of social housing blocks in Skovlunde, Ballerup near Copenhagen, in relation to an EU-Joule project, PV-VENT. The focus was on integration of PV modules for east–west oriented housing blocks, together with use of an optimal HRV system. Figure 4.49 is an illustration from the competition.

The winning entry was realised for a housing block with 27 apartments, where both decentralised HRV and centralised HRV systems were used. The result was that there were no problems with the decentralised HRV system, while the centralised HRV systems caused problems which took more than two years to resolve.

The problem was that there were complaints of a bad smell in a few apartments. It was investigated so see whether there was a problem with the HRV equipment. In the end it was found out that the only problem was failure in the regulation of the central HRV system by the installer, who had actually faked the regulation reports. This led to overpressure in some apartments and odours travelled through the stairway to other apartments.

The solar chimney in Figure 4.50, from Lundebjerg, Skovlunde, was the result of an architectural competition with a focus on PV and ventilation.

4.49
The winning entry from Tegnestuen Vandkunsten.

4.50
Ventilation chimneys gave
a position for PV modules
towards the south.

4.51
PV on gable and ventilation chimneys.

4.52
PV along windows towards
the east.

Peder Vejsig Pedersen and Katrine Vejsig Pedersen

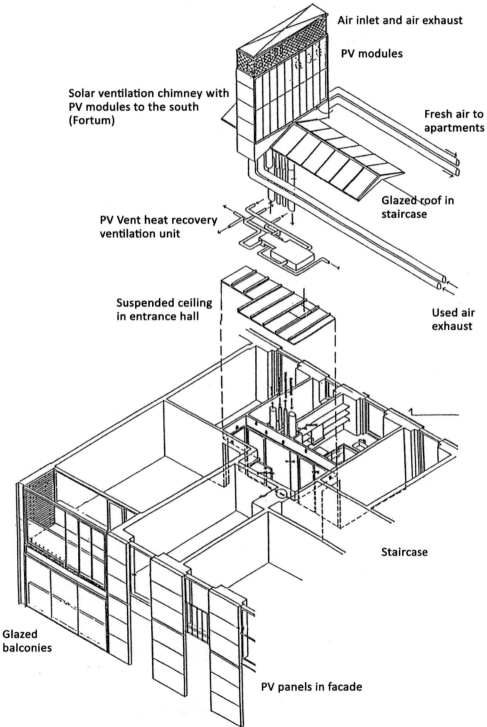

Air inlet and air exhaust

PV modules

Solar ventilation chimney with PV modules to the south (Fortum)

Fresh air to apartments

PV Vent heat recovery ventilation unit

Glazed roof in staircase

Suspended ceiling in entrance hall

Used air exhaust

Staircase

Glazed balconies

PV panels in facade

4.53
The principle of the PV-assisted ventilation design.

4.54
Ventilation duct integrated
in the entrance area of the
apartment.

Peder Vejsig Pedersen and Katrine Vejsig Pedersen

Munkesøgård, Roskilde

Around 2002 the Munkesøgård housing project in Trekroner, Roskilde was built with 100 housing units, making it the largest ecological housing project in Denmark.

The social housing part at Munkesøgård was made by a large contractor in a cheaper way than the family houses, but still with, for example, partition walls in clay. A 100 m^2 solar domestic hot water system was also used.

The family houses at Munkesøgård were based on the original architectural design and with a common house.

4.57
Social housing at Munkesøgaard was realised based on normal social housing building costs.

4.58
Unburned clay brick partition walls.

Peder Vejsig Pedersen and Katrine Vejsig Pedersen

4.59
Separation toilets.

Separation toilets were installed at Munkesøgård. Urine is sent to special storage tanks and used as fertiliser in connection to farming. It was planned to install composting toilets, but in the end the authorities did not allow this because it would not be possible for this to be handled by the maintenance staff in a safe way.

4.60
The Munkesøgård architect Martin Rubow on a site visit in Germany.

4.61
The 'Ecohouse' was launched as a low-energy house in Denmark based on experiences from Munkesøgård.

EHEN, the European Housing Ecology Network

4.62
EHEN EU brochure.

Peder Vejsig Pedersen and Katrine Vejsig Pedersen

Promotion material for the first Cenergia-initiated solar low-energy demonstration projects from the early 1990s included nine demonstration projects in seven EU countries with the EHEN, where €3 million of EU funding was utilised.

4.63
Solar low-energy housing in Swansea.

Building examples from the UK (Swansea, Hull and Milton Keynes) were realised in connection with the EHEN project in the 1990s. Solar collectors and heat recovery ventilation were something new, but the overall quality was not high.

4.64
The Hull housing project.

4.65
The Milton Keynes housing
project with Hastoe housing.

4.66
Portuguese bioclimatic
architecture in Sintra, with
built-in solar walls in the
EHEN EU project. TNUL
architects.

Peder Vejsig Pedersen and Katrine Vejsig Pedersen

European Green Cities project

A new EU low-energy demonstration project, European Green Cities, was realised from 1996 to 2000, again with €3 million of EU funding for ten demonstration projects in ten EU countries (www.europeangreencities.com).

4.67
Radstadt demonstration
project in Salzburg region.

The Austrian European Green Cities demonstration project was realised in Radstadt in the Salzburg region. It was realised in 1996 as the first larger solar low-energy housing in the Salzburg region. Many solutions were later copied in new projects, since this kind of building became the normal approach due to a unique energy-point financing system.

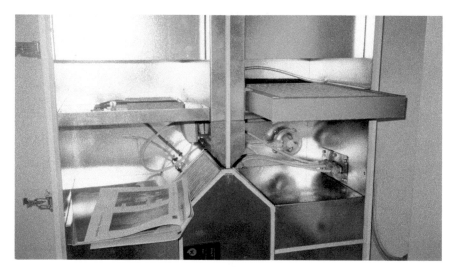

4.68
Example of the interior of a
heat recovery ventilation unit
in Salzburg, showing how
small HRV heat-exchangers
originally were, but most likely
with a quite limited thermal
efficiency.

4.69
European Green Cities study tour in Salzburg region.

The European Green Cities project was the start of many years of cooperation with the city and region of Salzburg in Austria. Figure 4.69 shows a site visit at a low-energy housing area where the heating supply was provided by a small solar heating and biomass-operated local district heating system.

4.70
Initiator and technical coordinator of the EU Ecobuilding target project, European Green Cities, Peder Vejsig Pedersen at the city hall in Copenhagen, signing the EU contract with Jens Frendrup from Green Cities, with the Mayor for the Environment overlooking.

Peder Vejsig Pedersen and Katrine Vejsig Pedersen

4.71
Italian project coordinator Salvatore Cali inspecting solar collectors in Brescia.

4.72
The Italian European Green Cities demonstration project incorporated use of large solar thermal collector systems for a low-energy housing project in Brescia.

4.73
Individual gas furnaces and
solar heated DHW tanks in
a Spanish demonstration
project.

Peder Vejsig Pedersen and Katrine Vejsig Pedersen

An EU-supported housing project in Villanova y la Geltru, south of Barcelona in Spain, utilised local solar-heated domestic hot water tanks with local gas heating units, which were placed on the balconies. The local pollution by gas burning seemed not to be a problem for the authorities. The European Green Cities partners were informed that results offering central heating solutions for whole housing blocks did not function well in Spain. This was documented in a previous EU–EHEN demonstration project (La Parra) in the city of Leida, where huge problems of organising this with individual heat meters led to the conclusion that it was always best with individual gas boilers. The truth was that due to the fact that mainly quite poor people lived in social housing in Spain, they will try to avoid paying for the heating bills and actually try to cheat. A solution could be to use the same type of prepayment meters that have been used in the UK (see the Newcastle example). It is a pity that central heating cannot be used, because we know that even in Leida it worked well with 100 per cent solar thermal in summer and 200 kWh/m² per year for heating on a yearly basis, mainly due to very little insulation. A primary problem is, however, that the gas company that operates the central heating system and the meters does not have any useful documentation. See also: http://dspace.unav.es/dspace/bitstream/10171/16808/1/RE_Vol%2026_01.pdf

JRC Ispra, Lago Maggiore, Italy

Cenergia won an international architectural competition concerning an energy-optimised renovation of the solar energy laboratory in JRC Ispra in Italy.

This was together with Metec Engineering (now Golder Associates) in Torino and the architect Piero Gatti, utilising a large Canadian solar wall installation

4.74
The facade to the north was optimised to allow daylight into the building.

together with thermal solar collectors. The project also incorporated large tropical fans to avoid high temperatures. Monitoring showed very good results for the project, with energy savings similar or better than calculated.

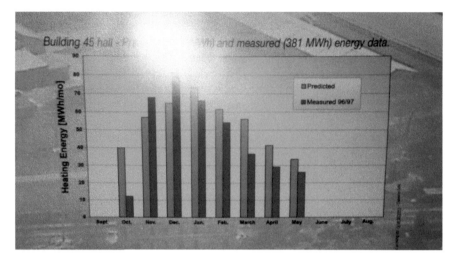

4.75
Good, monitored results were obtained.

4.76
Final design by Piero Gatti.

Peder Vejsig Pedersen and Katrine Vejsig Pedersen

4.77
The first drawings by Piero Gatti had a vision of using green roofs and facades, but later the Canadian solar wall was introduced and probably the main reason it was elected as the winner of the competition.

4.78
Solar solutions for the solar research hall in JRC Ispra incorporated the Canadian solar wall, preheating fresh air for the building and solar thermal collectors. Besides the Italian engineering company Metec from Torino (now part of Golder Associates), the project got support from Morten Pedersen from Calgary in Canada, who had many years of experience with optimised HVAC installations.

*Low-energy retrofit at housing block AAB Afd 23, Østerbro
in Copenhagen*

Here, 200 mm exterior insulation was allowed in the courtyard and also solar
heating faced the courtyard. The solar wall with transparent insulation is preheating
ventilation air for individual HRV systems and the solar collector is used for DHW.
The project was a finalist for the World Habitat Award 1995.

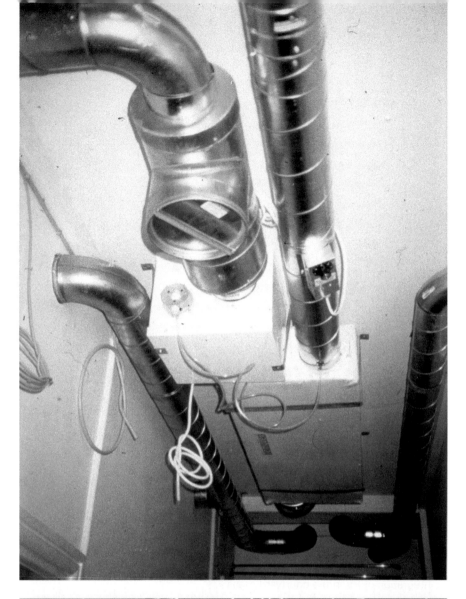

4.81
The solar low-energy retrofit
housing block at Østerbro
included individual HRV
systems in all apartments.

4.82
Katrine Vejsig Pedersen
visiting a building site at
Østerbro.

4.83
Close-up view of solar collectors at Østerbro.

Hedebygade housing renovation, Vesterbro, Copenhagen

The large-scale Hedebygade urban renewal project at Vesterbro in Copenhagen was part of the European Green Cities participation. It included several examples of integration of PV in the courtyard areas.

4.84
The Hedebygade project was the Danish European Green Cities demonstration project.

Peder Vejsig Pedersen and Katrine Vejsig Pedersen

4.85
Solar wall with a PV-assisted HRV system. Architects: C. F. Møller.

4.86
Example of PV integration in the Hedebygade urban renewal.

Frederiksberg, Lineagården housing retrofit with solar energy ventilation towers

4.87
EU-supported solar energy ventilation towers at Frederiksberg hiding the duct work for heat recovery ventilation and at the same time preheating fresh air.

4.88
Solar ventilation towers were implemented in 200 housing units at the Lineagården housing project at Frederiksberg in Copenhagen. Here they are also equipped with PV modules.

Hillsley Road, Portsmouth

The Hillsley Road low-energy houses in Portsmouth were part of the EU project Green Solar Regions. The overall result was not very good due to thermal bridges and air leakages. Balanced Danish HRV solutions were used here with good results.

Peder Vejsig Pedersen and Katrine Vejsig Pedersen

4.89
Solar low-energy housing at Hillsley Road, Portsmouth.

Stanhope Street, Newcastle

4.90
Stanhope Street housing area
in Newcastle with 350 housing
units.

4.91
Gas-driven CHP plant integrated at the Stanhope Street project.

In 1998 a substantial energy saving was obtained in the North British Housing Association EU-supported retrofit project with 350 housing units in Stanhope Street, Newcastle, UK. Good results were achieved by a local CHP plant with Danish district heating solutions with AVTB valves from Danfoss.

Use of a solution with a DHW heat-exchanger in each apartment was compared to use of individual DHW tanks. The result was a total energy use which was much higher in the housing blocks with DHW tanks due to much higher operational temperatures.

4.92
An Economy-7 electrical radiator: a heat storage radiator that was used before district heating was introduced.

Peder Vejsig Pedersen and Katrine Vejsig Pedersen

4.93
Stanhope Street before renovation.

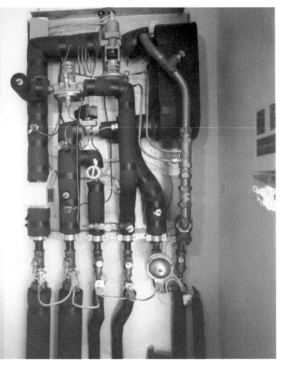

4.94
Danish district heating board with a DHW heat-exchanger.

4.95
Prepayment meters were used for all apartments.

A comparison of the heat consumption before and after the improvements is given in Table 4.1.

Table 4.1
Energy savings at Stanhope Street

| | **Heat Consumption** | | |
	before the renovation MWh	calculated in contract MWh	monitored MWh
Net heating demand	2,214	1,084	1,074
User effect	664	222	
DHW demand	537	369	717
DHW tank losses	369	185	180
Losses district heating		250	605
Total	3,784	2,110	2,576

There was good correlation between the actual heat consumption and the predictions concerning the space heating demand. A total saving of 32 per cent was achieved when compared to the heat consumption before the refurbishment.

Energy Signature

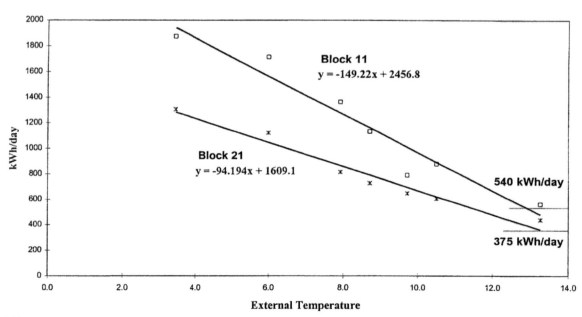

4.96
Monitored daily heat consumption based on monthly values is given as a function of the mean outside temperature.

The specific heat loss from the two blocks can be calculated based on a linear regression analysis, shown in Figure 4.96. The constant heat loss during the summer is shown in the table as the horizontal lines.

Table 4.2
Illustration of different heat use in two housing blocks of equal size

	Outside temperature °C	Heat Block 21 kWh	Heat Block 11 kWh
dec-96	3.5	39,662	59,970
jan-97	3.5	40,450	58,100
feb-97	6.0	31,350	48,000
mar-97	7.9	25,200	42,250
apr-97	8.7	21,800	33,900
may-97	10.5	18,750	27,200
jun-97	12.8	14,050	21,100
jul-97	16.2	12,100	18,000
aug-97	17.8	11,100	15,500
sep-97	13.3	13,150	16,850
oct-97	9.7	20,050	24,500
nov-97	8.2	25,101	36,996
		272,763	402,366

A linear regression analysis of the data in Table 4.2 gives the monthly heat consumption as a function of the external temperature in Figure 4.97. The total heat consumption in block 21 is 32 per cent lower than the heat consumption in block 11. Since the two building blocks are identical except for the domestic hot water heater, the higher heat demands in block 11 are caused by the domestic hot water tank compared with the heat-exchanger solution.

Figure 4.97 shows the return temperature from the three groups of blocks. In the medium block 25 per cent of the installations are with water tanks and in the low block 60 per cent of the installations are with storage tanks. The return temperature from the medium block is much lower than from the others, because the heat-exchanger installations can produce low temperatures closer to the cold feed water temperature.

Return Temperature
10-11 August 1997

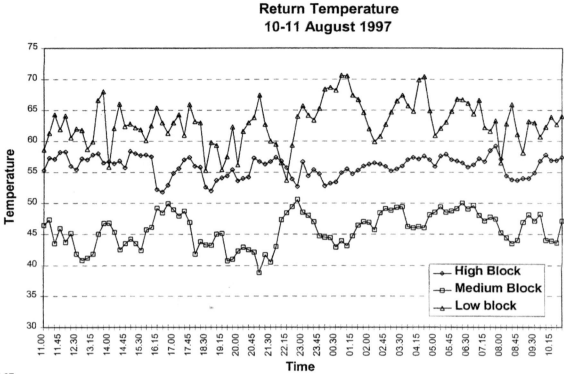

4.97
The return temperature from the three groups of blocks.

Block 11, Baxterwood Court

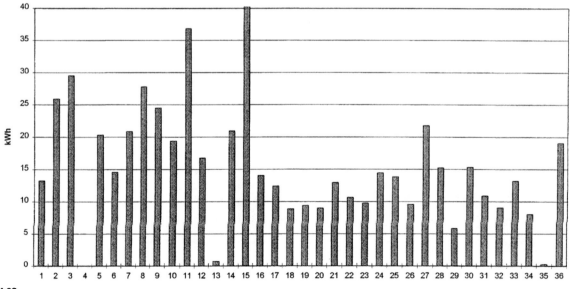

4.98
Monitored average heat consumption per day for 36 dwellings.

Peder Vejsig Pedersen and Katrine Vejsig Pedersen

More recent examples of low-energy building and renovation in Denmark

Low-energy housing blocks in Stenløse

The KAB low-energy Class 1/2015 housing project in Stenløse, realised in connection with the EU Concerto project Class1, www.class1.dk, is discussed here.

4.99
KAB prefabricated low-energy housing in Stenløse.

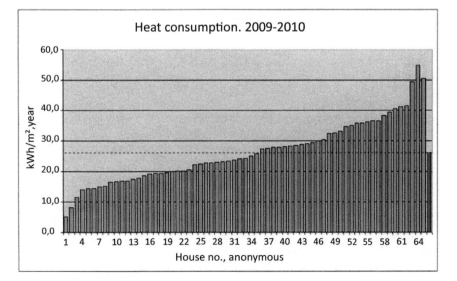

4.100
Yearly heat consumption in kWh/m² for the 65 housing units in order of consumption.

Yearly heating consumption in each of the 65 prefabricated housing units is shown in Figure 4.100. There is variation between 10 and 50 kWh/m² per year as a monitored value, with 26 kWh/m² as the average.

4.101
Ecovent heat recovery units installed in the prefabricated housing units.

Low-energy Class 1 housing project with 80 housing units at Ullerødbyen in Hillerød

4.102
Class 1 housing in Hillerød.

This project was realised in connection with the EU Concerto project SORCER, coordinated by the Danish engineering consultant company Cowi, which has conducted a very detailed monitoring of the operational results.

The 80 housing units were built in 2009 according to the low-energy Class 1/2015 standard in the new urban development area of Ullerødbyen. A large district heating system coupled 3,000 m² of solar thermal collector system had just been installed in front of the housing project by the local district heating company. In spite of this, the builder decided to introduce ventilation heat pumps as a basis for the heating, probably because of lower costs.

4.103
Arcon heating, solar collectors in Hillerød. Each solar collector is 12.5 m².

4.104
Large solar collector field for district heating in Ullerødbyen in Hillerød.

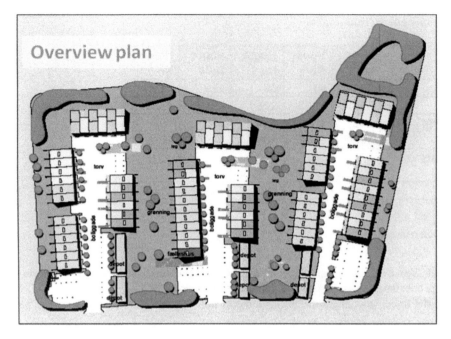

4.105
Housing overview.

Monitored data:

1. total electricity consumption (kWh)
2. total water consumption (m³)
3. PV production (kWh)
4. electricity consumption of heat pump unit (kWh)
5. electricity consumption of circulation pump (kWh)
6. total hot water consumption (m³)
7. heat production of heat pump for hot water (kWh)
8. heat production of heat pump for space heating (kWh)
9. electricity consumption auxiliary heater (kWh)
10. electricity consumption of ventilation unit (kWh)
11. room temperature (°C).

Unfortunately, it has turned out, it was not a good idea to use ventilation heat pumps because people were freezing and not happy about the installation. The monitored results are not impressive, but hopefully there is a lesson to be learnt from this. According to Danish law, a municipality cannot control the choice of heating system if the calculations are done in the right way.

Based on the total values shown in Table 4.3, and considering the area of the five monitored low-energy houses, the total annual primary energy consumption per square metre can be defined for each house. Furthermore, according to the Building Regulations BR08 the energy performance framework of a low-energy building Class 1 is:

Peder Vejsig Pedersen and Katrine Vejsig Pedersen

$(35 + 1100/A)$ kWh/m^2 per year,

where A is the heated floor area.

The energy performance framework and the measured/calculated total annual primary energy consumption for each low-energy house is then as shown in Table 4.4.

Table 4.3
Results for five housing units

	House 1	House 2	House 3	House 4	House 5
Area (m²)	40	61	68	115	115
Hot water (kWh)	1,845	2,085	1,668	6,936	4,334
Circulation pump (kWh)	442	452	424	696	707
Ventilation (kWh)	1,029	1,029	1,029	1,095	1,095
Space heating – weather corr. (kWh)	2,601	2,726	2,438	6,225	5,272
Total (kWh)	5,918	6,292	5,560	14,951	11,407

Table 4.4
Comparison of calculated and measured results for energy frame value and total energy requirements according to Be06 calculation, corrected Be06 calculation and according to measurements for small apartments. The connected values do not take optimal operation, like higher indoor temperature, into account

House no.	Area (m²)	Energy frame (kWh/m²/year)	Measured/calculated (kWh/m²/year)	Surplus (%)	Factor
1	40	63	148	137	2.4
2	61	53	103	94	1.9
3	68	51	82	60	1.6
4	115	45	130	192	2.9
5	115	45	99	123	2.2

The Rønnebækhave II passive housing project from 2005/2006

A shared ground-coupled heat pump was installed in the first passive house project in Denmark, Rønnebækhave II in Næstved. PV panels were used to provide the yearly supply of electricity in the heat pump.

Rønnebækhave II in Næstved comprises a small housing block with eight apartments built according to the German passive house principles.

4.106
Small, passive house housing block in Næstved in Denmark.

4.107
Example of passive house construction avoiding thermal bridges.

Peder Vejsig Pedersen and Katrine Vejsig Pedersen

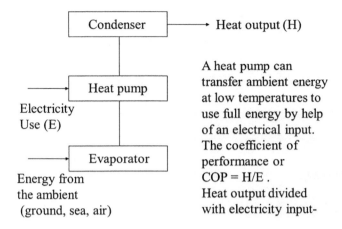

Condenser → Heat output (H)

Heat pump

Electricity
Use (E)

Evaporator

Energy from
the ambient
(ground, sea, air)

A heat pump can
transfer ambient energy
at low temperatures to
use full energy by help
of an electrical input.
The coefficient of
performance or
COP = H/E .
Heat output divided
with electricity input-

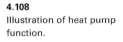

4.108
Illustration of heat pump
function.

Heat pump principle

The efficiency of a heat pump is dependent on the temperature levels of the
evaporator and the condenser, or rather the temperature difference between
the two. This is why ground-coupled heat pumps are quite efficient, especially
if they are combined with floor heating. The ground temperature is 8 °C as an
average mean, and floor heating can be provided by a forward temperature of
only 30–35 °C. In this case a yearly COP of 3–3.5 can normally be achieved.

4.109
The common heat pump for
eight housing units was made
with a buffer tank.

4.110
Building-integrated compact HRV.

Individual HRV units – the R200 from Ecovent – were placed in a small technical room. This enabled a 'dry' heat recovery efficiency of 85 per cent, together with a very low electricity use. Ecovent utilises an air-to-air heat-exchanger, which performs better than imported HRV heat-exchangers, which have 'dry' heat recovery efficiencies around 72–75 per cent.

Peder Vejsig Pedersen and Katrine Vejsig Pedersen

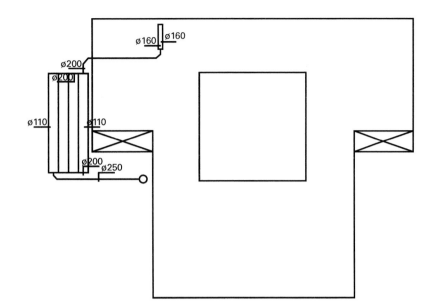

4.111
In two of the apartments
ventilation air was preheated
in the ground.

ø160 ø160
ø200
ø200
ø110 ø110
ø200 ø250

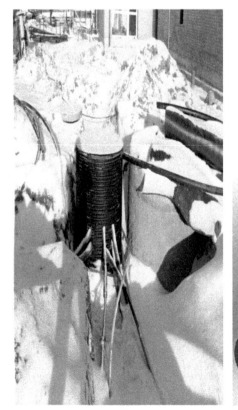

4.112
Pipes in the ground. A shared
IVT ground-coupled heat
pump (monitored COP of
3.2–3.4).

4.113
In order to meet passive
house room heating demand
of no more than 15 kWh/m²
per year it was necessary to
use low-energy windows from
Germany with a documented
U-value of 0.8 W/m²K.

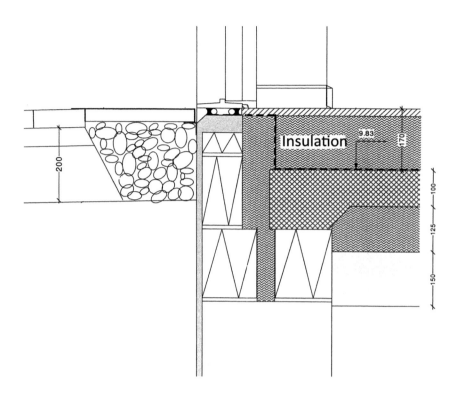

4.114
Construction without thermal bridges.

A low-temperature operation was used with floor and air heating. Thermal bridges are avoided and a BlowerDoor test showed good airtightness of 0.6 l/h at 50 Pa.

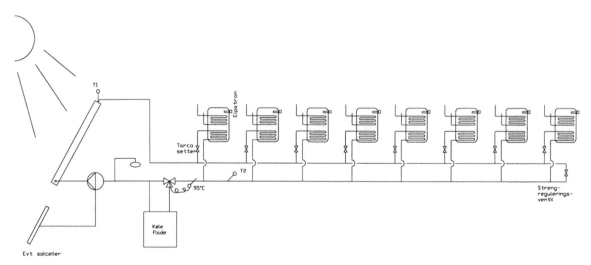

4.115
The apartments utilise a 28 m² shared solar DHW system from Arcon Solar Heating and individual 240 litre DHW tanks. The domestic hot water tanks have an electric supplement in the top. At higher temperatures than 95°C solar heat is sent to a ground-coupled heat-exchanger. The pump is supplied by PV electricity.

The experiences of the passive housing project in Næstved have been good, and the balanced HRV systems have been popular with the tenants. Two families claim they have avoided asthmatic symptoms since moving into the apartments.

Due to the limited knowledge concerning passive houses in Denmark, and the general policy concerning energy when Rønnebækhave II was built, it was difficult to get the project financed and to get funding for monitoring and follow-up. However, monitoring has been conducted since spring 2007, when an extra electricity meter was installed for the heat pump in order to document the actual COP value. The measurements documented that it was possible to obtain results almost according to the calculations, with an energy use for heating and DHW near 20–25 kWh/m^2 per year. Practical operation has overall been good, although a too-simple floor heating control unit had to be replaced with a Danfoss temperature monitoring system for the floor. Missing temperature controls of the solar DHW systems have also been mended.

The ground-coupled shared heat pump from IVT/Bosch is combined with air heating and floor heating. This is matched by PV electricity from a 5 kWp PV system in the roof on a yearly basis (Gaia Solar manufacture), so a CO_2-neutral design is obtained.

The energy use for the heat pump is the same as the yearly PV production of the PV modules, which is 4,000 kWh or 6.3 kWh/m^2 per year.

4.116
50 m^2 of PV modules are integrated in the roof facing south, in connection to a solar thermal collector area of 28 m^2 for DHW.

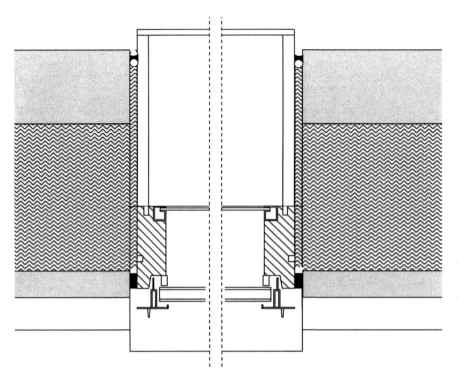

4.117
Construction without thermal bridges. Horizontal section. The link between the rockwool insulation and the insulated part of the window insulation is unbroken and thermal bridges will therefore be avoided.

Energy frame value for passive house dwellings in Rønnebækhave II in Næstved, DK compared to old and new Building Regulations.

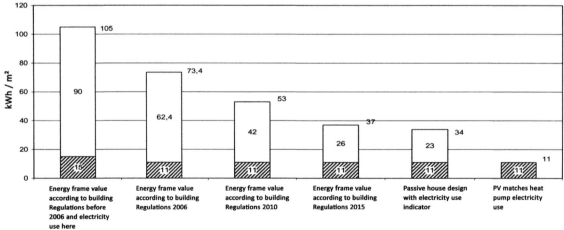

4.118
Energy frame value reaching zero-energy operation for heat supply.

Active House renovation in Albertslund

Due to foreseen large-scale renovation of 3,000 housing units in the city of Albertslund, west of Copenhagen, it was decided that the municipality would only accept a renovation standard equal to the minimum standard for new buildings, but the aim is actually to reach a much better standard, aiming towards the building standards in Denmark for 2015 or 2020. Also, the possibility of meeting a 100 per cent zero-energy standard based on solar energy combined heat and power, which can be financed by the local district heating company, is being investigated. At the same time this will support the aim to cover 10–15 per cent of all electricity use in Albertslund with solar power by 2020.

The first Danish Zero Energy Housing Renovation project – The Hyldespjældet test house in Albertslund with the prefabricated installation element on the roof, the Solar Prism, and with prefabricated construction elements – was realised in 2009 in a public–private partnership also involving the large building component companies, Velux, Rockwool and Danfoss, together with the companies Kuben Management, Moe & Brødsgaard, Rubow Architects and Cenergia, in cooperation with the Danish Technological Institute and the housing association, BO-VEST. This has supported the choice of Albertslund as the Energy Municipality 2011 in the five Nordic countries.

Purpose and approach

The EU Building Directive 'nearly zero-energy' standard for new buildings by 2020 has implications for existing buildings, energy supply systems and energy quality.

A basis for a completely new way to build in Europe has been created by the recast of the EU Energy Performance for Buildings Directive from 2010, which demands that new buildings should achieve a nearly zero-energy standard based on local renewable energy sources by 2018 for new public buildings and by 2020 for all new buildings, at the same time as the EU members states shall develop incentives to ensure a similar development for existing buildings.

Denmark is well on the way to meeting these goals due to recent improved energy-saving demands in the building regulations, including new protected low-energy classes for 2015 and 2020.

But when it comes to which energy supply solutions to use there are a lot of discussions.

At the present, Denmark has the highest utilisation in the world of district heating, often in combination with combined heat and power and waste incineration, covering more than 50 per cent of all heating demand; due to this it seems like an obvious choice to rely on these technologies for the low-energy buildings of the future.

In general it is important that the district heating sector is able to deliver adapted energy-optimised solutions for low-energy buildings of the future, so it is possible for municipalities to utilise this in new urban developments. This should also include a policy for use of solar heating as a renewable energy source, which will have the best economy in large, centralised solar heating systems.

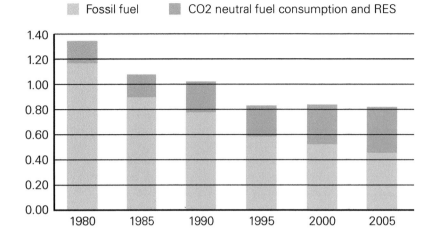

Fossil fuel CO2 neutral fuel consumption and RES

4.119
The amount of fossil fuel has decreased over the years in district heating in Denmark.

Based on 50 years of development in Denmark, district heating is a very reliable technology which is already by all accounts the most sustainable energy supply solution. And even more is planned for – e.g. the larger Copenhagen area, in which by 2025 the district heating will be 100 per cent CO_2-neutral, based on a huge utilisation of biomass.

In connection to this it is also important to work on ideas of how the zero-energy building of the future should look, including a view on how far the energy use needs to be reduced when introducing photovoltaics or solar cells to ensure the zero-energy standards can be met by use of local renewable energy sources.

Active House Alliance

The Active House Alliance was first established in spring 2010 and was later made official at an international Active House conference in Brussels in April 2011, where new 'Active House Specifications' were agreed.

For Active House Class 1 buildings all energy is produced from renewable energy sources, either integrated into the building or taken from the nearby collective energy system and electricity grid. Active House Class 2 buildings are defined in the same way, but only for the operational energy use, as electric appliances are not included here. Active House Class 3 and 4 buildings, according to the Active House Standard, have an energy use of 15 kWh/m² per year and 30 kWh/m² per year, respectively, which can be useful in renovation projects (www.activehouse.info).

According to the EU Commission Building Directive, all new buildings need to be nearly zero-energy by 2020, and EU member states need to show how a similar quality can be obtained for existing buildings as well. The Active House Specification shows how this can be done.

It is important to note that there is a demand in the Active House Specification for energy monitoring, verification and follow-up. This is new

Peder Vejsig Pedersen and Katrine Vejsig Pedersen

4.120
Active House Specifications
and promotion.

compared to the situation in Denmark, where focus is on good calculation procedures, but with no link to what the actual energy use will be in practice in realised building projects.

This poses a good possibility to introduce the same demands for 'verification' of all new building projects within a two-year period, which has already been introduced in Sweden. In Denmark there are no rules of responsibility concerning the actual energy quality in building projects, and this is having a major influence on how much focus consultants and contractors have on the realised energy quality.

One area with an actual focus on energy quality in Denmark is the area of airtightness of constructions. This can be controlled quite easily; due to this, and a five-year guarantee period for building failures, there is now consensus on the importance of living up to this demand in the Building Regulations.

In many cases the concept of zero-energy building depends on the fact that it is possible to calculate energy frame values by adding heating use together with electricity use, multiplied by a factor of 2.5.

When designing a low-energy building with a low-energy frame value, this can be matched on a yearly basis with a similar contribution from renewable energy sources, which will typically be a mix of solar electricity from PV panels and solar heating.

For many people it is easy to understand how to use a heat pump, which uses electricity and matches the yearly electricity use by solar electricity from PV panels to reach a zero-energy design. But it is more difficult to understand how the same can be achieved in a district heating area. But based on the use of the mentioned energy frame values it is actually quite easy to design a zero-energy house based on the use of district heating.

A new paradigm for renovation

Investigations by the Building Research Institute in Denmark have shown that even though very costly renovation of social housing has been realised every year by funding from the Danish Social Housing Fund, only a mere energy saving of 10 per cent has been obtained in practice. Due to this it is necessary to find a completely new way of organising renovation of social housing, so possible high energy-saving prospects are realised in practice in a new and much more cost-effective way. This includes a focus on a public–private partnership to obtain good results in practice, e.g. based on:

- smart prefabricated facade and roof solutions with a high insulation value, airtightness and without thermal bridges;
- smart decentralised comfort systems with built-in highly efficient heat recovery and very little electricity use for the fans;
- reduction of domestic hot water use by an increased focus on saving water;
- use of cost-efficient and architectonic optimised 'solar energy roofs' in combination with cost-efficient roof renovation concepts;
- introduction of a continuous registration of energy savings, including contribution from renewable energy solutions.

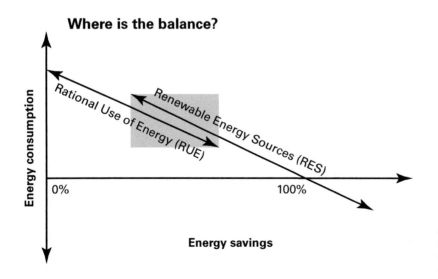

4.121
How much do you need to save and how much solar energy do you need?

Peder Vejsig Pedersen and Katrine Vejsig Pedersen

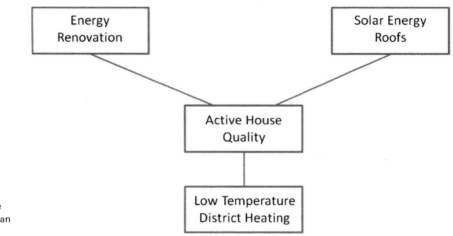

In connection to building renovation it is possible to use the same energy frame values which are used for new-builds; use of renewable energy solutions like solar energy systems can be taken into account. In order to live up to the new low-energy 2020 class in Denmark or even a zero-energy or plus-energy building standard, renewable energy becomes very important. In connection to this it is relevant to utilise the Active House Standard, which has clear definitions of yearly energy balance in different classes and which includes demands for indoor air climate quality as well as for documentation of performance results and follow-up.

Research and technical development work in connection to large-scale renovation in Albertslund

Background concerning housing renovation in Albertslund

The municipality of Albertslund has agreed to become the Danish climate test site concerning energy-efficient renovation of concrete housing areas, since many billions of DKK will be invested during the coming years. In order to show that advanced energy solutions that live up to the Danish low-energy class 2015 or 2020 levels are possible to realise with a positive economy and improved comfort, it has been, with supplementing funding from the Danish EUDP programme, an ambition to establish 'example' renovation projects in several places in Albertslund as a means to identify the extra cost and total economy of these solutions. In Albertslund South the largest retrofitting project in Danish history is currently in progress. The project is called 'Master Plan South'; it includes the retrofitting of about 2,200 dwellings, and it consists of three phases:

1. 623 multi-storey dwellings – the construction work was initiated in 2007;
2. 552 terraced houses – the construction work was initiated in 2013;
3. 1,000 atrium houses – the construction work was initiated in 2014.

The Danish Housing Association, BO-VEST, is responsible for this, the largest and most costly renovation plan for social housing in Denmark. BO-VEST represents the housing societies that own the 2,200 dwellings (Albertslund Boligselskab and Vridsløselille Andelsboligforening). At BO-VEST, and in the municipality of Albertslund, there is a real devotion to optimising the renovation approach, including a low-energy renovation design with improved indoor air climate and an optimised energy supply solution. The occupants of the 2,200 dwellings are also eager to reach an ambitious level of energy efficiency when the dwellings are renovated.

Main results

The first demonstration project in Albertslund, which was supported by the Danish EUDP programme, the Social Housing Fund and Albertslund Municipality, was made in 2009 in the housing area of Hyldespjældet.

The basic idea was to utilise prefabricated construction elements, together with a prefabricated installation module on the flat roof – the Solar Prism. The Solar Prism is a modular system developed by Cenergia and Rubow Architects in cooperation with Velux and Danfoss. In Hyldespjældet it includes all installation elements, so it is not necessary to introduce installations inside the renovated dwellings. With good financing based on municipal guaranteed loans for the social housing investments, this secures a reasonable balance for the tenants, between extra yearly capital costs and the value of the energy savings, so high-quality up-to-zero-energy housing renovation can be utilised both in coming large-scale renovation projects in Albertslund as well as in similar projects in other parts of Denmark. At the same time the concept has huge utilisation prospects in Europe. In 2012 the Hyldespjældet demonstration project led to a contribution of the Energy Globe Award 2012 for Denmark.

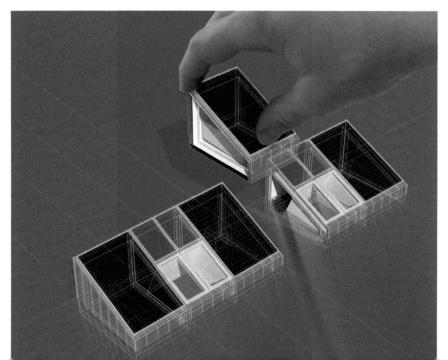

4.123
Illustration of the modular Solar Prism, here presented by Velux. Design: Martin Rubow and Peder Vejsig Pedersen.

4.124
Prefabricated Solar Prism
and how it was utilised in
Albertslund with a heat
pump combined with a
solar domestic hot water
tank and an HRV system.
Facing the sun, the Velux roof
windows and solar thermal
collectors work together with
PV modules. Design: Martin
Rubow and Peder Vejsig
Pedersen.

This project was in 2011 followed up by a demonstration project with six
terraced houses in different energy qualities designed by Nova 5 Architects
and Niras engineers, and during 2012 two privately owned houses were also
renovated as new demonstration projects in Albertslund.

The Solar Prism combines different prefabricated modules according to
individual needs.

Albertslund South terraced houses present economy for different solutions
compared to the cost of integrated zero-energy renovation concepts.

Other initiatives are focusing on how to introduce an 'Active House'
renovation standard and on how to use solar energy combined heat and power
as a new and cost-effective approach that interacts with the existing combined
heat and power-based district heating in an optimised way, where solar heating
is directly fed into the combined heat and power-based district heating network.
At the same time PV electricity is fed into the electricity grid in a 1:1 ratio, just
like the existing CHP plants that are operating for the large-scale district heating
system in Copenhagen.

4.125
The Solar Prism is placed on
the flat roof and includes all
installation elements.

4.126
At Hyldespjældet the
first zero-energy housing
renovation in Denmark
introducing prefabricated
construction elements from
Rockwool and Solar Prism
installation elements.

4.127
A decentralised HRV system in
each apartment with building-
integrated components,
low electricity use and very
satisfied users. The same
technology was used in the
Solar Prism in Albertslund.

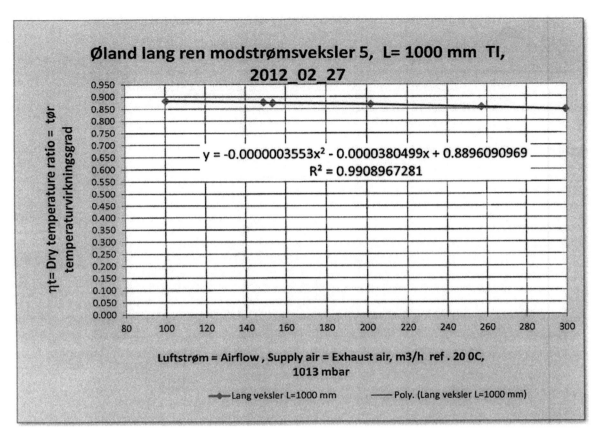

Øland lang ren modstrømsveksler 5, L= 1000 mm TI, 2012_02_27

$\eta t=$ Dry temperature ratio = tør temperaturvirkningsgrad

$$y = -0.0000003553x^2 - 0.0000380499x + 0.8896090969$$
$$R^2 = 0.9908967281$$

Luftstrøm = Airflow , Supply air = Exhaust air, m3/h ref . 20 0C, 1013 mbar

◆— Lang veksler L=1000 mm ——— Poly. (Lang veksler L=1000 mm)

4.128
Test of compact HRV unit from Ecovent/Øland showing dry heat recovery ventilation efficiency greater than 85 per cent. This technology is used in the Solar Prism in Albertslund.

4.129
Test terraced houses completed in different low-energy standards by summer 2011.

4.130
PV and Velux roof window integration in a typical one-family house from the 1960s at Flintager 55 in Albertslund. Architect: Martin Rubow.

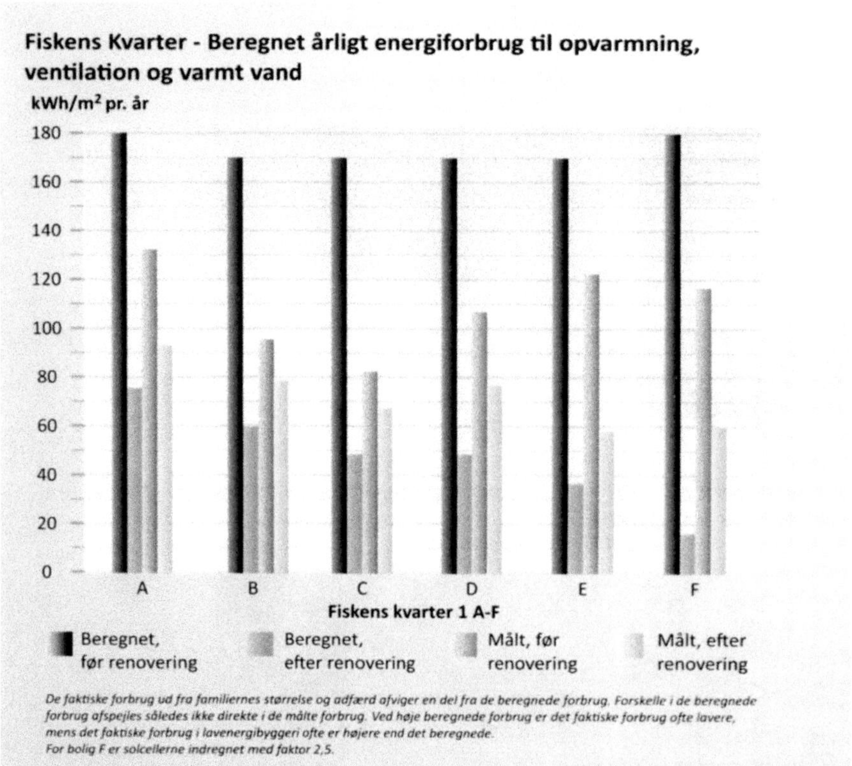

Fiskens Kvarter - Beregnet årligt energiforbrug til opvarmning, ventilation og varmt vand

kWh/m² pr. år

[Bar chart with x-axis labeled A, B, C, D, E, F. Y-axis from 0 to 180.]

Fiskens kvarter 1 A-F

■ Beregnet, før renovering   ■ Beregnet, efter renovering   ▌Målt, før renovering   ▌Målt, efter renovering

De faktiske forbrug ud fra familiernes størrelse og adfærd afviger en del fra de beregnede forbrug. Forskelle i de beregnede forbrug afspejles således ikke direkte i de målte forbrug. Ved høje beregnede forbrug er det faktiske forbrug ofte lavere, mens det faktiske forbrug i lavenergibyggeri ofte er højere end det beregnede.
For bolig F er solcellerne indregnet med faktor 2,5.

4.131
Comparison of calculated and measured energy use in kWh/m² per year for heating, ventilation and domestic hot water for six different renovated housing units with an increased low-energy standard from A to F. The project is situated in 'Fiskens Kvarter' in Albertslund. Dark and light grey columns show calculated and measured heating and hot water use before renovation, while dark and light green columns show calculated and measured heating use after renovation. Project supported by the Danish EUDP programme.

Peder Vejsig Pedersen and Katrine Vejsig Pedersen

4.132
The single-family housing full-scale renovation project in Flintager 55 in Albertslund demonstrates integration of PV modules and Velux windows.

4.133
A zero-energy housing renovation project with prefabricated construction elements and Solar Prism. Hyldespjældet in Albertslund.

Example of low-cost energy renovation of an old house outside the district heating area

A combination of new windows, little insulation and airtightness improvements, use of low-cost HRV and solar DHW, together with an air-based heat pump, led to an energy standard like new-build housing in Denmark. With the addition of 8 kWp of PV, an almost zero-energy solution is obtained.

4.134
Inverter for an 8 kWp PV system with roof- and ground-based PV.

4.135
Proposed integration of PV panels in relation to renovation of an old house from 1925 in Sorø at Zeeland. The house will be finalised in spring 2014.

4.136
Insulation of a high cellar below the kitchen.

4.137
A special crawl space insulation design with blown-in paper granulate insulation with glue ensures a 100–150 mm insulation and airtightness of the floors.

4.138
Air-based heat pump installation in a cellar with a heat pump and buffer tank, together with DHW and a buffer tank for solar DHW heating system.

4.139
Energy-efficient air-based heat pump from Danfoss integrated near facade and with short pipe lengths to installations in the cellar.

Gyldenrisparken social housing retrofit, Copenhagen

The Gyldenrisparken housing estate at Amager near Copenhagen consists of a number of four-storey concrete housing blocks built between 1965 and 1969, with 450 apartments in all. Work on optimising the renovation was made in the EU project Demohouse (www.demohouse.net).

Calculations were made for several scenarios using the national Danish Be06 calculation tool. The five investigated scenarios were:

1. The existing situation.
2. Standard renovation (facades insulated with extra 100 mm mineral wool and new windows with U-value of 1.4).

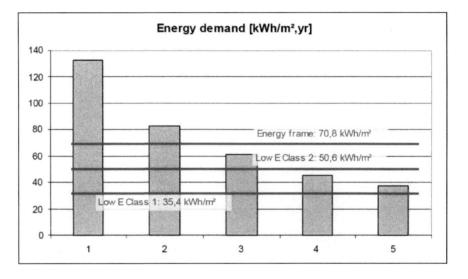

4.142
Calculated energy use for each of the five scenarios in Gyldenrisparken. Low-energy Class 2 is now the Building Regulation 2010 standard and low-energy Class 1 is equal to the low-energy class 2015 standard.

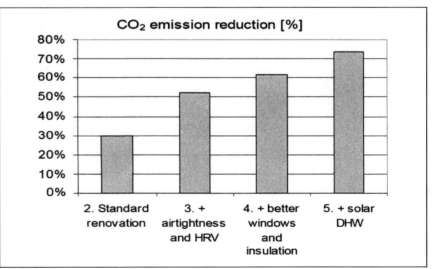

4.143
CO_2 emission reduction.

3. Standard renovation including airtightness and balanced HRV (infiltration at 0.1 ACH and heat recovery efficiency of 85 per cent).
4. Scenario 3 with improved windows (U-value 1.0) and an extra 100 mm mineral wool in the roof.
5. Scenario 4 with solar DHW. 75 m^2 of solar collectors will supply approximately 60 per cent of the annual DHW consumption.

Problems encountered

Use of a new type of low-cost HRV system ensures the highest energy saving and best economy for the user, together with an improved indoor air climate. To ensure low costs and a high tenant-satisfaction level, practical testing in some of the apartments was performed prior to realising a whole housing block with HRV solutions.

The total economy calculation confirmed that the above-mentioned energy-saving measures were the most interesting and that it should actually be possible to reach a new low-energy building 2010 quality with an acceptable economy for the users. It was also confirmed that use of low-cost HRV systems was very important to meet this target.

4.144
Alternative HRV systems integration.

Peder Vejsig Pedersen and Katrine Vejsig Pedersen

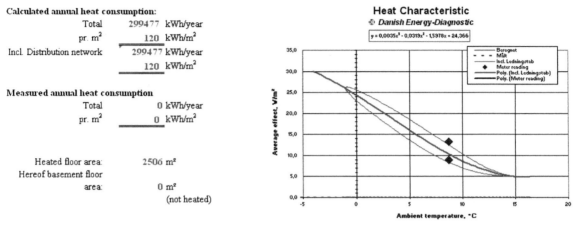

Programme for evaluation of energy performance and diagnostic of operation malfunctions
© Danish Energy-Diagnostic
Input in blue cells
Output in red cells   **Building:** Gyldenrisparken | Current consumption |

Calculated annual heat consumption:
Total   299477 kWh/year
pr. m²   120 kWh/m²
Incl. Distribution network   299477 kWh/year
120 kWh/m²

Measured annual heat consumption
Total   0 kWh/year
pr. m²   0 kWh/m²

Heated floor area: 2506 m²
Hereof basement floor
area: 0 m²
(not heated)

Heat Characteristic
© Danish Energy-Diagnostic
$y = 0,0035x^3 - 0,0319x^2 - 1,5378x + 24,366$

4.145
Example of energy signature calculation.

Scenario	Costs, operation heating [Euro]	Costs, operation electricity [Euro]	Costs, operation water [Euro]	Costs, operation total [Euro]	Costs, investment [Euro]	Pay-back time [years]
1. Existing situation	26,349	8,640	10,282	45,271	0	0
2. Standard renovation	16,370	8,640	9,768	34,777	345.942	33
3. + airtightness and HRV	11,229	5,760	7,711	24,700	433.142	9
4. + better windows and insulation	8,064	5,760	7,711	21,535	628.373	21
5. + solar DHW and PV for ventilation	6,451	5,040	7,711	19,202	675.873	21

4.146
Costs and pay-back time for the five different energy-saving scenarios.

Pay-back time for scenario 3 with HRV and airtightness is 87,200/10,077 = 9 years, while for scenario 4 with better insulation and windows it is 282,431/13,242 = 21 years. For an energy Class 1 solution with solar DHW and PV the pay-back time is 47,500/2,333 = 21 years (2008 analysis of renovation at Gyldenrisparken housing project in Copenhagen).

4.147
The renovation of the facades in Gyldenrisparken has unfortunately resulted in many thermal bridges.

In connection to a successful use of decentralised HRV systems, different types of ventilation duct integration were tested in 27 apartments in the Gyldenrisparken renovation project, where a total of 450 apartments were renovated.

4.148
Building-integrated heat recovery ventilation ducting was tested with success in multi-storey apartments in Gyldenrisparken in Copenhagen in 2008. The HRV unit is integrated with a suspended ceiling in the bathroom.

4.149
Circular ducting should be avoided, but a rectangular solution was successful.

The five-year investigation of the HRV systems was successful. People were pleased and no problems had occurred.

4.150
Even though a test apartment was made prior to the renovation, it was primarily to test the production plan. Avoidance of thermal bridges was not part of the checking work.

Demonstration of improved ventilation with heat recovery

One of the most important tasks in the Green Solar Cities EU Concerto project has been to improve the quality of HRV technology as stated in the contract. HRV is considered to be the most important technology to secure a low-energy standard for both new buildings and retrofit building projects to meet the improved low-energy class 2015 and 2020 standards in Denmark, at the same time securing a good indoor air climate.

An important barrier to achieving good results, however, has been that the so-called dry heat recovery efficiency of the existing HRV systems for the Danish and European market is quite low, with typical values of around 70–75 per cent, while the electricity use at the same time is quite high; this means it is difficult for most HRV systems to meet SFP values of less than $1,200\,J/m^3$ in electricity use for individual systems and $2,100\,J/m^3$ for larger, common systems.

Due to this situation, RTD work in connection to optimised and low-cost HRV technology with high heat-exchanger efficiency and low electricity use has taken place in Denmark in cooperation with the Danish ventilation companies in relation to the Green Solar Cities project.

By March 2011 it was concluded that it has been possible to reach very good results concerning this, and at the same time it has also been possible to present good prospects concerning a completely new low-cost HRV technology which especially is very useful for retrofit building schemes.

Heat recovery efficiencies as high as 90–93 per cent were confirmed by use of an advanced calculation tool from the Technological Institute, and more importantly this has also been confirmed by practical testing of the technology. For a long time the performed RTD work unfortunately documented problems of obtaining a high airtightness of the HRV heat-exchanger in practice. In the end these problems were solved, however.

It is possible to give examples of several practical HRV tests with more than 90 per cent HRV efficiency and with an extremely low electricity use. For example, researchers from the University of Alborg have shown SFP values around 300–$400\,J/m^3$ in connection with retrofit housing in practice, which is only 33 per cent of the Building Regulation demands. According to a test in a single-family house, a total electricity use for both fans of only 13 W was monitored compared to normal values of 35–50 W.

In total these improvements alone can secure an extra saving of $11\,kWh/m^2$ per year in the energy frame value, which is actually equal to 25 per cent of the low-energy class 2015 standard. To be able to document the very low electricity use in practice, RTD work has also been carried out to secure a continuous survey of this by a low-cost internet connection for the users.

Finally, it can be stated that the RTD work performed on the HRV systems mentioned has been of great importance for the project, and for the first large-scale integration in Langgadehus in Valby good results have been obtained.

Based on this it has been possible to document this high-performance technology in several demonstration projects connected to the Concerto project at Valby and to document through energy surveys that the targeted results are obtained in practice, including documentation of the possibility of using very limited amounts of PV module areas of 1–2 m² per housing unit to cover the yearly electricity use.

Use of outstanding HRV technology in the EU Concerto project in Langgadehus in Valby

In Langgadehus in Valby it was possible to use the new HRV technology with up to 90 per cent dry efficiency for the HRV heat-exchanger, together with a very low pressure loss.

4.151
Prefabricated rooftop dwellings in Langgadehus in Valby built according to the Danish low-energy Class 1 standard.

4.152
In Langgadehus the Ecovent Eco 450 model was installed in 2010.

Detailed monitoring of the HRV systems in Langgadehus has been conducted in cooperation with the Danish Building Research Institute, including experiments with means of demand-controlled ventilation. In connection with this, remote monitoring and an option for control via the internet have recorded the electricity use in some of the apartments. Intelligent survey and control equipment has been introduced as an option for online survey and monitoring in several cases.

The HRV technology used has also been tested at VIA University College in Denmark, showing HRV efficiency over 90 per cent. Remote control and survey of HRV systems was made by Zense Home equipment.

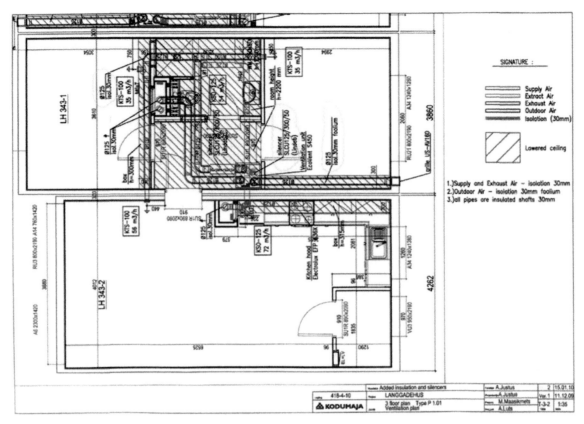

4.153
HRV integration in Langgadehus.

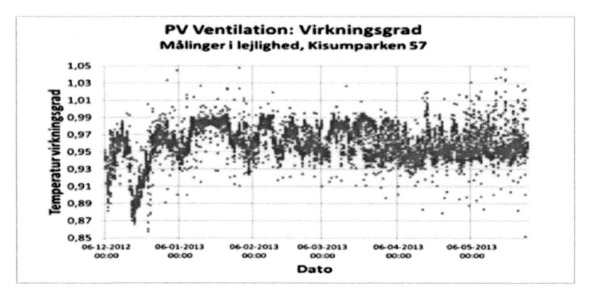

4.154
Test results by VIA University College in Denmark showing HRV efficiencies of 92–98 per cent in actual HRV installation from December to May (95 per cent as a mean).

Peder Vejsig Pedersen and Katrine Vejsig Pedersen

4.155
The test set-up at the VIA
University College.

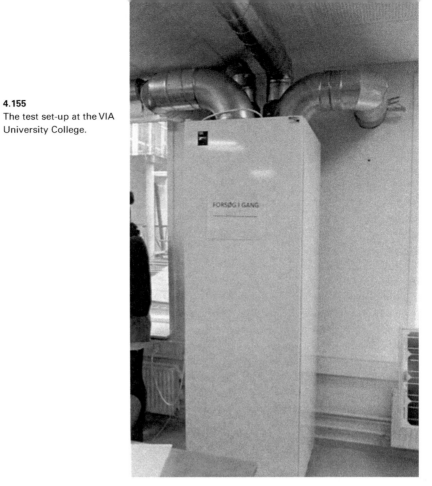

4.156
Monitored HRV efficiency
at VIA University College
showing efficiencies between
90 and 95 per cent. In this
'black box' test it is not the dry
HRV efficiency that is tested.
This will typically be 5 per
cent lower when good fans
are used.

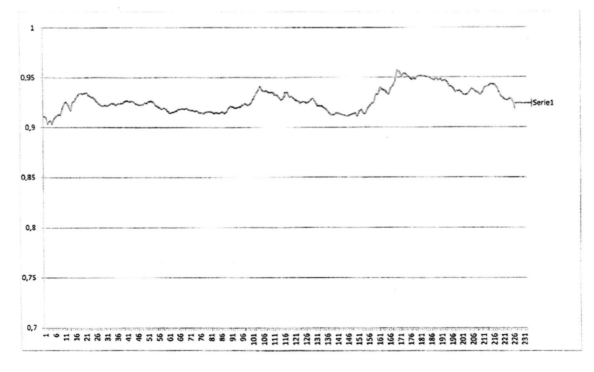

Centralised HRV systems

In the Hornemannsvænge large-scale housing renovation in Valby, a very high electricity use was documented for the centralised HRV systems covering 24 apartments for each system: 1,100 kWh of electricity per apartment per year was used for ventilation, leading to yearly costs of around €320.

In most cases good central HRV systems for new-build housing can be designed with around half of this electricity use, which is still high, since the best decentralised ventilation systems will only use around 200 kWh per year of electricity per apartment.

From a cost point of view there is a more simple service of centralised HRV systems, so total yearly operational costs for the mentioned system in Hornemannsvænge will be approximately €350. The operational cost of decentralised HRV systems will typically be around one-third of this, including filter exchange, which is more complicated when it involves service people. A new automatic filter exchange box from Ecovent could change this in the future, since it only needs to be serviced once every ten years.

Decentralised HRV systems

Principle solution B
Type Apartment
Area of the dwelling
Kitchen 12 m2 + Bathroom 3 m2 + Rooms 70 m2 = Total 85 m2

Room height 2,5m and occupants 2

Unit
Ventilation with heat recovery
Heating: district heating or electricity
CO2/temperature/humidity controller

Description
There is installed one decentralized unit in the center of the apartment which provide the rooms with fresh preheated air.

The ventilation flow in each of the rooms is mechanical controlled by the CO2 content/temperature/humidity in the room. But it is also possible for the user to control the system manually.

There is inlet in all rooms except the bathroom and kitchen where there is placed exhaust outlets.

This solution only needs one or two holes in the façade in order to get the fresh air supply and remove the exhaust air.

Requirements

SEL > 0,6 kJ/m3 (BR10 > 1,0 kJ/m3)

Airtightness > 1,0 l/s m2 at 50 Pa

Noise > 27 db(a)

Ventilation
BR10
Minimum air change rate 0,3 l/s m2 or 126 m3/h which should be possible to increased to 0,41 l/s m2 (Exhaust air from kitchen on 20 l/s and bathroom 15 l/s)

4.157
HRV unit under a suspended ceiling in the entrance room of an apartment, taking fresh air from the facade and leading used air to the facade or roof.

20111104

1905

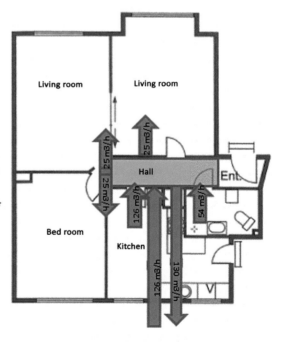

The focus here is on two types of decentralised HRV systems. The most common solution is the decentralised HRV system with one HRV unit per apartment, where the HRV unit is placed under the bathroom ceiling. This solution can lead to a 28 per cent energy saving compared to a standard renovation.

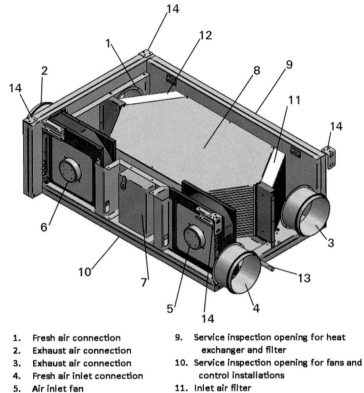

4.158
CAD drawings of the Ecovent renovation model with the same HRV technology as in Langgadehus. The HRV is typically installed below the ceiling of the bathroom.

1. Fresh air connection
2. Exhaust air connection
3. Exhaust air connection
4. Fresh air inlet connection
5. Air inlet fan
6. Exhaust air fan
7. Control box
8. Counterflow heat exchanger

9. Service inspection opening for heat exchanger and filter
10. Service inspection opening for fans and control installations
11. Inlet air filter
12. Exhaust air filter
13. Condensation water container and extract
14. Ceiling mounting point

Size: 1065 mm x 640 mm x 330 mm

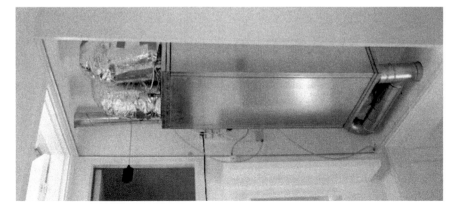

4.159
Photo of HRV installation under the loft in an apartment.

As an alternative it is possible to install an HRV system which is room-based, with typical airflow volume of 40–80 m³/h. In Valby a rooftop extension has been made with a innovative low-cost decentralised HRV design.

In most cases two room-based units will be enough, e.g. leading fresh air also to the neighbouring room, and connected to a moisture-controlled exhaust in the bathroom, which seldom operates. This can be combined with a recirculation cooker hood in the kitchen.

New building-integrated room-based decentralised HRV solutions

4.160
One-family house in Valby before renovation.

4.161
A single-family house in Valby with a rooftop extension with innovative HRV technology. PV panels are used to match the yearly electricity use for ventilation.

Heat recovery ventilation unit with radiator function

To be able to overcome the other main barrier for the HRV technology in retrofit building projects, which are the extra costs, important RTD work on the so-called HRV radiator model has been carried out. The idea is to substitute a normal radiator with an integrated HRV radiator unit, which has the same size and which can contribute with heating as well as ventilation.

The main need for this technology is low or no noise, easy integration, a compact design as well as low cost. Noiseless prototype systems have been identified, together with an electricity use as low as 5 W at air volumes around 60–80 m³/hour.

4.162
Window-integrated HRV unit with radiator function.

4.163
Window-integrated HRV unit and normal radiator.

Window-integrated HRV was established at the P. Knudsens Gade housing area, which is part of the Concerto area of the Green Solar Cities project in Denmark. The window-integrated HRV module was placed beside a normal radiator, which is used for heat accounting purposes. The HRV module has a radiator function as well and can heat to either room temperature or higher temperatures.

4.164
The new windows for the test apartment were supplied with built-in fresh air intake and used air exhaust integrated in the bottom of the window equal to 140 mm diameter.

4.165
The normal exhaust ventilation from the bathroom is controlled with an extra damper, so only a high moisture content or 5–10 minute operation based on a user push of a button will operate the exhaust ventilation.

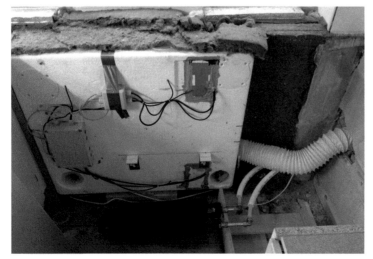

4.166
A compact windowsill-integrated low-cost HRV unit and a built-in heating convector as well as an extra inlet air function without heating connected to the neighbour bedroom.

A built-in thermostat controls the inlet air to the kitchen, so at least room-temperature air is always secured. With two HRV units each of these will provide 50–60 m³/h of fresh 90 per cent preheated air for the quite small test apartment.

4.167
The cooker hood in the kitchen is changed to a recirculation cooker hood for simplicity.

4.168
Drilled holes for injecting paper granulate insulation.

When balanced heat recovery ventilation is provided to an old apartment it is important to secure a good airtightness of the apartment. The main problem is to avoid air leakages through the old wooden deck construction. In the test apartment this is done by introducing non-toxic paper granulate insulation near the facade. The round hole in the floor shows one of the spots for this injection.

Examples of PV projects in buildings

PV example projects in Copenhagen

4.169
The canteen at the architectural school in Copenhagen before renovation.

4.170
PV in a new glazed construction in the roof.

Peder Vejsig Pedersen and Katrine Vejsig Pedersen

A good example of a project with building integration of PV modules is the renovation of the old canteen at the Architectural School in Copenhagen. Part of the PV modules also secured PV-assisted ventilation to avoid overheating in sunny periods.

4.171
A couple of the PV elements were used to secure PV-operated extra ventilation against overheating with 1,500 m³/h of airflow.

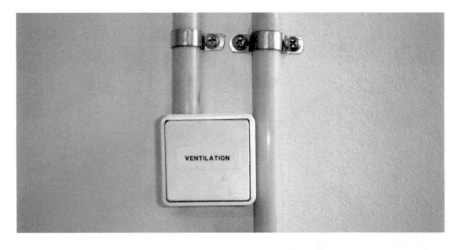

4.172
PV-assisted exhaust ventilation.

4.173
PV-assisted ventilation for a non-heated glazed housing extension, where temperatures up to 50 °C in sunny periods were reduced by more than 10 °C.

Copenhagen PV Co-op (www.solcellelauget.dk)

4.174
In relation to the
EU-Resurgence project,
Erik Christiansen from EBO
Consult and Peder Vejsig
Pedersen from Cenergia
established the first PV Co-op
in Copenhagen in Denmark in
2004, where people bought
shares in a 40 kWp system
on top of a municipal-owned
building.

BIPV at the Museo di Bambini in Rome

4.175
An EU-supported demonstration project in the late 1990s
gave support to building integration of PV modules in the
Museo di Bambini in Rome.

4.176
Sun shading by use of PV

4.177
Interior of Musei de Bambini in Rome

BIPV in Boston

The Artists for Humanity charity helps young people to work with their artistic skills instead of creating trouble in the streets. Their headquarters in Boston is LEED certified and uses PV in the roof. Boston was one of the first cities to develop a solar map for the city, where good locations for solar energy could be identified.

4.178
PV mounted on top of the roof at Artists for Humanity Epicenter in Boston.

4.179
View of the Artists for Humanity building in Boston.

4.180
PV demonstration project in Boston, USA.

PV demonstration site in Lausanne, Switzerland

4.181
In Lausanne, Switzerland,
PV example projects have
been distributed around the
area of a normal parking lot
at the local university (www.
demosite.ch).

Frederiksberg, Copenhagen, Denmark

4.182
At the Lauritz Sørensens
Gård housing renovation,
PV panels were integrated
with a Canadian solar wall
roofing installation, preheating
ventilation air before reaching
the heat recovery ventilation
units.

4.183
Example of PV integration in
an old tile roof at Trekanten in
Frederiksberg, Copenhagen.

4.184
Early example of PV
integration in urban renewal
projects in Copenhagen.

Netherlands

In 2002 the 2 MWp PV installation at the Floriade exhibition was for a short time
the biggest in the world.

4.185
2 MWp PV installation in
Holland from 2002. The
biggest in the world at that
time.

4.186
PV on a flat roof.

4.187
Example project with PV roofs.
Integration of PV modules.
Designed by Bear Architects.

From 1998 to 2006 many good examples of PV installations on flat roofs were made in the Netherlands. Figure 4.187 shows a project connected to the EU-Resurgence activities.

United Kingdom

4.188
Peabody Trust PV in London.

PV projects were being implemented by the Peabody Housing Trust housing association in London around 2006 in connection to the EU-Resurgence project. The original idea was to develop an alternative concept for all future roofing retrofit projects.

4.189
PV at Bedzed, UK.

The Bedzed housing project in the south of London was made by the Peabody Trust just after 2000 and was based on a unique architectural sustainable building solution, never seen before and actually very appealing to many people. At a site visit in 2005 in connection to the European PV demonstration project Resurgence, led by the Peabody Trust, the partners were informed very openly about all of the problems in the project.

Nevertheless, as an example of a holistic approach that seeks to combine sustainability, energy efficiency and use of renewables, the project had a very strong communication effect, at least in the UK.

4.190
In Bedzed PV modules were integrated in an architectural design.

Austria

4.191
The largest passive house building in Vienna in Austria also has a large number of PV panels integrated.

PV integration examples from 2012

Examples of PV integration are available from 2012, with rapid expansion of PV technology in Danish housing units (from 12 to 500 MWp in one year).

4.192
Typical integration of PV modules here in Måløv, without much architectural consideration and with the modules lying a bit higher than the roof.

4.193
An even worse integration of PV around Velux windows. It is clear that a roofing system, which can include PV, is missing.

4.194
Another unoptimised PV design.

4.195
A slate roof with PV integrated in a nice way.

Søndergården nursery home in Måløv

4.196
PV at Søndergården nursing home in Måløv.

With an aim to build according to low-energy class 2020 at the Søndergården nursing home in Måløv, both PV modules and a large ground-coupled heat pump system were installed. Late in the process a conflict of combining ventilation with heat recovery equipment and PV modules on the flat roofs became evident

4.197
PV and ventilation equipment on the roof.

The PV systems are not well integrated in the buildings. A better solution would have been to install decentralised HRV systems and use the roof for PV production.

4.198
An unoptimised PV solution.

Figure 4.198 shows a poor combination of PV systems and ventilation systems on the roof. In the EU Concerto project in Valby the same problem was seen in low-energy class 2015 buildings. But when aiming at a 2020 standard, better solutions are necessary, such as decentralised ventilation systems, which typically use half as much electricity compared to centralised ventilation systems.

4.199
Example of the new Danish hourly net metering scheme for PV.

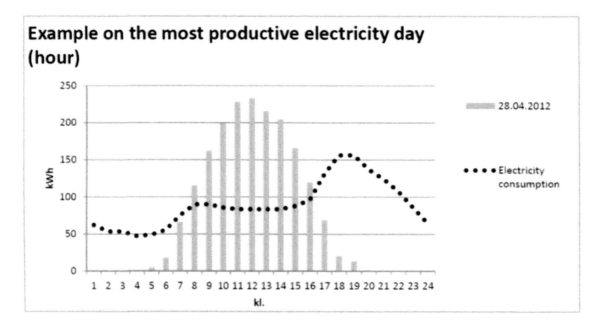

Example on the most productive electricity day (hour)

28.04.2012

●●●● Electricity consumption

Peder Vejsig Pedersen and Katrine Vejsig Pedersen

When PV is consumed directly in the hour it is used, you get the same energy saving value as the normal electricity costs.

When PV electricity is not used directly, it is sold to the grid at a considerably lower cost.

PV-assisted ventilation at the Tjørnehøj School in Brøndby near Copenhagen

4.200
PV modules facing south are integrated with a ventilation tower and a low-electricity consumption DC fan that exhausts air from the glazed room to avoid overheating.

4.201
PV modules in the glazed room create sun shading.

4.202
A PV-operated exhaust fan
uses electricity from one of
the PV modules to exhaust
hot air.

Solengen

Solengen in Hillerød is a low-energy shared-ownership housing project with partition wall-integrated HRV systems. The use of electricity was matched by PV at the common house.

4.203
Sarnafil roof with built-in PV
installation at the common
house in Solengen, Hillerød.

4.204
HRV integrated in the partition
wall between bathroom and
entrance room in Solengen.

*Building-integrated PV roofing system at Søpassagen housing
block in Copenhagen*

4.205
Søpassagen housing block
before integration of PV
modules.

4.206
The so-called 'Copenhagen'
roof consists of a flat roof
part and a sloping roof part. PV
panels on the top of the flat
roof part are placed so they
cannot be seen from the street
area.

Peder Vejsig Pedersen and Katrine Vejsig Pedersen

4.207
PV panels integrated in
the sloping roof part in
Søpassagen are chosen as an
official PV example project by
the Copenhagen municipality.

4.208
Karin Kappel, Solar City Copenhagen, and Klaus Boyer Rasmussen, Solarplan, discussing the Søpassagen project.

4.209
View of a large part of the 45 kWp PV installation at Søpassagen.

4.210
The PV installation at Søpassagen is combined with a roof terrace for the tenants, overlooking the lakes in the centre of Copenhagen.

x

Sustainable and energy-efficient urban planning in Malmø

The district heating system for the Western Harbour area in Malmø is based on large heat pumps taking energy out of the ground water situated in underground aquifers. The so-called ATES (aquifer thermal energy storage) system with cold and hot boreholes is here used as the basis for both cooling needs in summer and heating needs in winter. There are 2,500 m² of thermal solar collectors covering 15 per cent of the annual heating demand. PV systems cover part of the electricity use for the heat pumps.

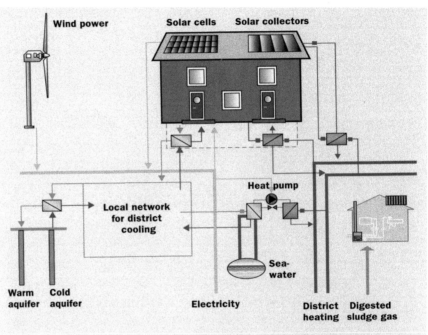

4.211
In the Western Harbour in Malmø a so-called ATES is used.

4.212
European Green Cities meeting at Western Harbour in 2001.

4.213
Example of a solar thermal
solution on the facade of
the very good ecological
restaurant 'Salt and Brygge'
in the Western Harbour in
Malmø.

Peder Vejsig Pedersen and Katrine Vejsig Pedersen

4.214
Example of PV installation
with shading function in the
Western Harbour in Malmø.

4.215
Example of large PV
installation on an existing
housing block in Malmø.

4.216
The landmark building in the
Western Harbour in Malmø is
the 'Turning Torso', which is
also based on a sustainable
building approach.

The Western Harbour area in Malmø is the most ambitious sustainable development area you can visit in the Copenhagen region. The basis was an international sustainable urban development exhibition held in 2001 (Bo01); very detailed planning here secured a very holistic sustainable energy and building design.

Peder Vejsig Pedersen and Katrine Vejsig Pedersen

4.219
Part of Malmö City maintenance centre in Augustenborg with PV sun shading. Faeces from seagulls proved to be a problem because you have to clean the PV modules at least once per year.

At the Augustenborg area in Malmø there have been experiments with 'green roof' solutions at municipal-owned buildings for more than ten years, used as a basis for successful local implementation against rainwater problems in the area. Also here are experiments with PV and solar thermal.

4.220
An American-inspired shredder solution in one of the kitchens in the 'Turning Torso' building makes it possible to collect biological waste in a simple way through the waste water system.

4.221
The storage for biological waste from the 'Turning Torso' building in the Western Harbour in Malmø. This is collected by a lorry every 14 days and transported to the biogas plant at the Northern Harbour, where it is used for CHP production.

4.222
An example of innovative housing design in the Western Harbour area, with vertical gardens in the back of the housing block.

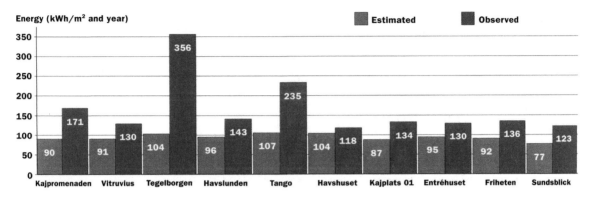

4.223

Comparison of calculated and measured energy use in kWh/m² per year in low-energy housing projects in the Western Harbour area in Malmø. Up to 60 per cent higher energy use than was calculated is seen here.

The main reasons for the typical 20–60 per cent higher measured energy use for heating (Figure 4.223) is that airtightness and avoidance of cold bridges is not good enough; control systems are not adapted to the limited heat use; ventilation systems are not controlled well enough and heat recovery of ventilation is not functioning as calculated; floor heating is often used; hot water use and indoor temperature are higher than expected; and only 70 per cent of electrical use is a benefit for heating.

These results were an important reason in Sweden to put responsibility on the builder to provide performance documentation within two years of construction.

Chapter 5

Visions

PEDER VEJSIG PEDERSEN

This chapter on 'visions' includes a focus on zero-energy building, active houses and performance documentation, together with a view of the needed development of active roofs and facades of the future.

Development of zero-energy buildings

At the same time as the demands of the Building Regulations are increasing all over Europe due to the EU Building Directive, it is also clear that there is a missing link between this and the widespread and increasing market for renewable energy, like solar energy. This is not so strange when you look at the large-scale PV and solar thermal plants being implemented, but when it comes to the installation of solar energy systems on buildings it is actually a negative situation when you try to fulfil quite demanding low-energy building requirements in both new-build and large renovation projects, as these solutions are only seldom considered,

Due to this situation, the effects of which could be seen at the time of the launch of the Valby PV plan in Copenhagen, Peder Vejsig Pedersen from Cenergia made the choice to try to develop a small zero-energy housing unit in 2003, when there was funding available from the large EU PV demonstration programme Resurgence. It was probably some of the best money ever spent because the positive feedback was enormous compared to the quite small investment.

Zero-energy test housing unit at Toftegårds Plads in Valby

The realisation in 2003 of a high-quality zero-energy housing unit insulated with paper granulate and with building-integrated PV in the roof was a useful way to brand the combination of low-energy building and use of PV. This project created the background for establishing the Solar City Copenhagen Cooperation and the prefabricated housing unit, SOLTAG. The support of Copenhagen municipality, and a long-term supporter of solar energy in the Danish Energy Agency in the form of Mr Jens Windeleff, was crucial to this success.

5.1
The small zero-energy test house was developed by Cenergia in 2003 and exhibited in Valby in Copenhagen.

5.2
Erection of the zero-energy test house.

Peder Vejsig Pedersen

5.3
The small zero-energy housing
unit in Valby was made in a
diffuse, open, wooden form,
without membranes and with
the use of paper granulate
insulation, which would
ensure moisture would never
collect in the construction.

5.4
A metallic roof with Unisolar
thin-film PV panels.

SOLTAG CO₂-neutral rooftop apartment

In 2004 a new zero-energy housing unit was developed with Martin Rubow as the architect and with support from the City of Copenhagen, the Danish EFP programme and EU funding. This was the SOLTAG CO_2-neutral rooftop apartment, which was made in cooperation with Velux, Cenergia and Kuben Management. The costs were ten times higher than the Toftegårds Plads house and due to a very tight time schedule, different problems like thermal bridges could not be avoided. But as a promotion of the idea of zero-energy building it was a very strong initiative, and also led to the Danish Energy Saving prize in 2005.

5.5
The CO_2-neutral rooftop apartment was exhibited in Ørestaden in Copenhagen for a year before it was moved to the premises of Velux (www.soltag.net).

5.6
Based on the good results with the SOLTAG CO_2-neutral rooftop apartment, the Velux group developed a similar solution for hotter climates, which was exhibited in Bilbao in 2006.

SOLTAG summerhouse at Gudmindrup Beach in Denmark

5.7

This summerhouse was built in 2007 in Gudmindrup Beach in Denmark. It was designed by Martin Rubow, who also designed the SOLTAG rooftop apartment which was developed in cooperation with Velux in 2005. The roof is a combination of metal cladding with PV modules, Velux roof windows and solar thermal solar collectors that works together with a heat pump system.

5.8

Velux roof windows facing south and north secure good daylight quality. The house is insulated with 300 mm paper granulate and has heat recovery ventilation and floor heating.

5.9
The interior, showing a good daylight function.

5.10
The first heat pump design included an air-based heat pump which took heat in from the roof surface. Since it was impossible to avoid a high noise level, this was later changed to a ground-coupled heat pump.

5.11
The backyard has been dug out to make a place for pipes in the ground for the heat pump.

5.12
Wooden floor construction is seen here, with paper granulate insulation and floor heating elements securing a low-temperature heating system.

5.13
To the left, the low-energy rafter that prevents thermal bridges in the overall construction. The whole house was constructed from this construction element. Piping from solar collectors is coming from the roof.

5.14
Ekofiber paper granulate insulation was blown into the construction, which was made with an open membrane design allowing possible moisture to dissipate to the surroundings.

Active House Standard

Based on the above experiences of development of zero-energy housing, including new initiatives by Velux to develop a series of model homes, an initiative to develop basic principles for 'Active Houses' was developed, with support from the building industry, and was officially launched at an Active House conference in 2010 in Brussels.

At the same time actual Active House specifications were developed, which can be found at www.activehouse.info. Here there is defined a number of specifications within areas like *energy*, *indoor climate* and *environment*. In the energy area there is a focus on the following areas: energy balance, energy design, energy supply, energy monitoring and verification and follow-up.

The proposed basis for evaluation of the Green Solar Cities project based on Active House Specifications is shown in Figures 5.16–5.21. Sources and tools for this information are: national calculation methodology, national primary factors and climate data, the EU Directive on energy performance of buildings (2010/31/EC of 19 May 2010), EU Directive on the promotion of the use of energy from renewable sources (2009/28/EC of April 2009), and national test and evaluation methods, like the BlowerDoor test and thermograph photos.

5.15
Active House Radar showing performance concerning energy, environment and indoor climate.

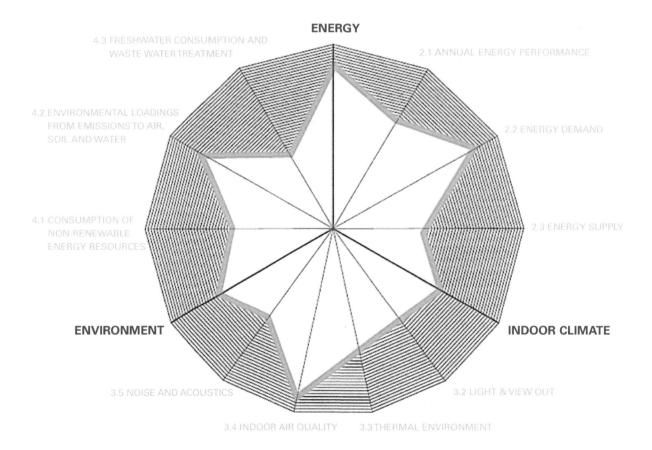

PARAMETER	EVALUATION METHOD AND CRITERIA
Quantitative	
Energy and CO_2-calculation	Calculation of primary energy and CO_2-emissions shall be based on the national calculation methodology, using nationally adopted efficiency/conversion and emission factors, as well as climate data. The definition of the heated floor area shall follow the national definition.
Annual energy performance	The annual energy performance shall be based on primary energy calculations and it includes calculation of the energy demand of the building, the energy demand of appliances and a calculation of the renewable energy being used.
	An Active House is classified according to the annual primary energy use, where:
	1: $\leq$ 0 kWh/m2 for the building and appliances 2: $\leq$ 0 kWh/m2 for the building 3: $\leq$ 15 kWh/m2 for the building 4: $\leq$ 30 kWh/m2 for the building (modernization)

5.16
Annual energy
performance.

PARAMETER	EVALUATION METHOD AND CRITERIA
Quantitative	
Annual energy demand for the building	The annual energy demand includes energy demand for space heating, water heating, ventilation, air conditioning including cooling, technical installations and electricity for lighting.
	1: $\leq$ 30 kWh/m2 2: $\leq$ 50 kWh/m2 3: $\leq$ 80 kWh/m2 4: $\leq$ 120 kWh/m2 (modernization only)
Annual energy demand for appliances	The annual energy demand for appliances includes white goods, television, computers and similar equipment.
Demand to individual products and construction elements	The requirements to individual products and construction elements (i.e. minimum thermal resistances, maximum thermal bridge effects, and air tightness) shall at least follow requirements set in national building regulations.
Qualitative	
Building management system	An Active House should be prepared for an easy and user-friendly control of the indoor climate and the energy use in the building.
Demand to individual products, construction elements and appliances	Have the chosen products and construction solutions been evaluated from a cost perspective and maintenance view and how was the decision about the individual products and construction solutions taken?
	Have the best energy performing solutions for appliances been chosen?
Architectural design solutions	How are architectural design solutions used to reach a holistic approach of the building, as well as to reach a low energy demand?

5.17
Energy demand.

PARAMETER	EVALUATION METHOD AND CRITERIA
Quantitative — Annual energy supply	The annual energy supply from renewable energy and CO_2-free energy sources shall be calculated and divided into the different sources (PV, Wind, Heat pumps, Solar Thermal, Biomass, etc).
Origin of energy supply	The renewable energy sources can either be used on the building, the site, in a nearby energy system or electricity grid. The energy supply can be a mix of the above and follow the classifications below: 1: 100% of the energy is produced on the plot 2: more than 50% of the energy is produced on the plot 3: more than 25% of the energy is produced on the plot 4: less than 25% of the energy is produced on the plot
Sources of renewable energy	The definition of renewable energy sources follows the EU Directive on the promotion of the use of energy from renewable sources (2009/28/EC of 23 April 2009).
Performance of renewable energy system	Requirement to performance of the individual renewable source shall follow the national requirement in building legislation. As an alternative to national requirements the requirement in the EU Directive on the promotion of the use of energy from renewable sources (2009/28/EC of 23 April 2009) can be used.
Qualitative — Design	How have you worked with integration of renewable energy as a part of the building design and typology on the building and the plot?
Origin of energy supply	Has the energy supply been evaluated from a cost perspective and how was the decision about the origin of the energy supply taken?

5.18
Energy supply.

PARAMETER	EVALUATION METHOD AND CRITERIA
Quantitative	
Specification of figures	The calculation of the annual demand shall specify the energy demand for the individual areas, as specified in section 1.2: • space heating, water heating, ventilation, cooling and air-conditioning, electricity for technical equipment, electricity for lighting and electricity for appliances The annual energy performance evaluation shall specify the supply from: • individual renewable energy sources integrated into the building • energy supply from local energy system and the share of renewable energy as well as the CO_2 emissions from the local energy system
On site control	In order to prove that the build energy solutions meet the designed level, an on site control of the used solutions and products must be established by a certified expert. The air permeability of the building and the thermal bridges must be evaluated during the construction phase.
Monitoring	The energy used and the energy produced must be monitored on a yearly basis. Metering devices are to be used for all types of energy production/consumption at the building level.
Qualitative	
Quality control	What kinds of quality control of the energy performance have taken place, where and when was it done?
Commissioning	How will the commissioning of the building take place and will the commissioning include user behaviour, number of users, heating and ventilation system, control of dynamic solutions and production of renewable energy? If the use of energy is different from the calculated values, what kinds of initiatives are planned to meet the calculated values?
User guidelines	What kinds of initiatives have been taken to secure that the house owner and users of the building have the relevant information about the expected performance of the building and guidance to use and optimize the building?
Energy control	What kinds of initiatives have been taken to secure the users' possibilities to control and optimize the use of energy?
Maintenance	What kinds of initiatives have been taken to secure the users a possibility to maintain the technical equipment as well as other parts of the building that affect the energy performance?

5.19
Energy validation.

PARAMETER	CRITERIA AND EVALUATION METHOD
Quantitative	
Standard fresh air supply	Fresh air supply can be evaluated by looking at CO_2-concentrations indoor at room level during occupancy. CO_2 is a good indicator for the amount of bio effluents, pollutants from humans, in the air. Hourly values and maximum levels of CO_2-concentration preferably are determined with a dynamic simulation tool. Assuming standard occupancy rates (e.g. two persons in a master bedroom) and standard CO_2-production per person assumptions.
	The limit values for indoor CO_2-concentration in living rooms, bed rooms, study rooms and other rooms with people as the dominant source and that are occupied for prolonged periods are:
	1. 350 ppm above outdoor CO_2-concentration
	2. 500 ppm above outdoor CO_2-concentration
	3. 800 ppm above outdoor CO_2-concentration
	4. 1100 ppm above outdoor CO_2-concentration
	Reference: EN 15251: 2007. Note that this standard also gives the fresh air supply rates (e.g. for living rooms and bedrooms) in l/s/m2 that are needed to comply with the CO_2-requirements above.
Minimum fresh air supply	When dwellings are unoccupied a minimum air change rate of 0.2 ACH should be maintained, to remove pollution from material emissions, appliance emissions, etc.
Dampness	In rooms with periodic damp production peaks (esp. kitchens, bathrooms and toilets) sufficient extraction must be guaranteed to avoid dampness and mould problems. The minimum exhaust air flow in these 'wet rooms' should be as specified in national building codes or guidelines. If these are not available, see EN 15251: 2007 for example design values (in 3 categories) for exhaust air flows in kitchens, bathrooms and toilets.
	Dampness problems furthermore are prevented by a building envelop that is well insulated and free of cold bridges. In order to prevent internal or surface condensation that could lead to mould growth and a deterioration of the air quality inside. Comply, also in renovation project, as much as possible with local building requirements for thermal insulation and e.g. temperature factor.
Qualitative	
Individual control	It should be possible to manually influence the air change rate in the rooms (especially living room, kitchen and bed rooms), e.g. by opening windows.
	In case mechanical ventilation is installed, it should be possible to adjust the airflow rate at 3 or more levels. Additionally, the ventilation may be demand-controlled based with CO_2 or humidity sensors.
Low-emitting building materials	Building components and materials (e.g. processed wood products, paints and sealants) should be evaluated with regards to chemicals emitted from them, and low-emitting components are preferred.
	Preferably use Indoor climate labeled materials. For example materials with the Danish Indoor Climate label, the Finnish M1 label, the German AgBB or GUT label or the French AFFSET label.
User instruction	In case of complicated ventilation systems or unusual user restrictions (e.g. concerning interior materials that can be introduced by the occupants) an easy to understand 'indoor air user instruction' should be provided. This document should explain how systems work and what is expected of end-users (e.g. as far as operation of systems and maintenance is concerned).

5.20
Indoor air quality.

	PARAMETER	CRITERIA AND EVALUATION METHOD
Qualitative	Building traditions	How is the design of the building reflecting a relationship with the regional building traditions? E.g. regional materials, architectural typology and handcraft is analyzed and used as design parameters?
	Climate	How is the design of the building adapting to the potentials and constrains of the local climate? E.g. creating private outdoor spaces with a comfortable climate and access to sunlight that encourage to healthy active outdoor living?
	Street- and landscapes	How does the design impact on existing street- and landscapes? E.g. provision for children to play safely outside the house and supporting the public outdoor space for local behavior, needs and tradition?
	Infrastructure	How is the infrastructure supporting a healthy, comfortable and ecological transportation? E.g. connection and distance to nearest public transport for commuting, distance to school and supermarket and the possibility of easy and safe use of bicycles?
	Ecology and land use	How is the building optimizing the relationship with the local ecology and land use and at the same time minimizing environmental risks? E.g. maximizing surface for seepage of rainwater, minimizing use of land, preserving fauna.
	Climate changes	How are possible risks caused by climate changes (storms and flooding) identified and limited in the design of the building and landscape?

5.21
Cultural and ecological context (sources and tools: ISO 14040. ISO 14025. CEN TC 350).

There is a need for a new approach to energy-efficient housing renovation showing how renewables can be introduced, together with an optimised energy supply. The Active House Specification can be useful here.

In the Active House Specifications there is a demand for energy monitoring, verification and follow-up. This is a new approach compared to the situation in Denmark today, where there is a lot of focus on good calculation procedures, but, like in most other countries, no link to the actual energy use in practice in realised building projects.

A good possibility here could be to introduce the same demands for 'verification' of all new building projects within a two-year period, which has already been introduced in Sweden.

First Active House in Canada

5.22
The Great Gulf Homes Active House is the first Active House in Canada. The house is situated in Thorold in the Niagara Region of Ontario, Canada.

5.23
The Scandinavian-inspired Active House is built with prefabricated construction elements from the Brockport factory in Toronto, as is the case for the neighbouring houses, in spite of their Victorian appearance.

5.24
The neighbouring houses
are the normal standard of
'Victorian'-inspired houses,
which are very popular in
Canada.

5.25
Inspection of the prefabricated
construction in the cellar
during a site visit.

5.26
A Velux solar domestic hot water tank is installed in the basement of the house. The house uses an air-conditioning system with air heating in the heating season, like most Canadian one-family houses.

5.27
The Great Gulf Homes director, Tad Putera, explains details of the Active House in Thorold.

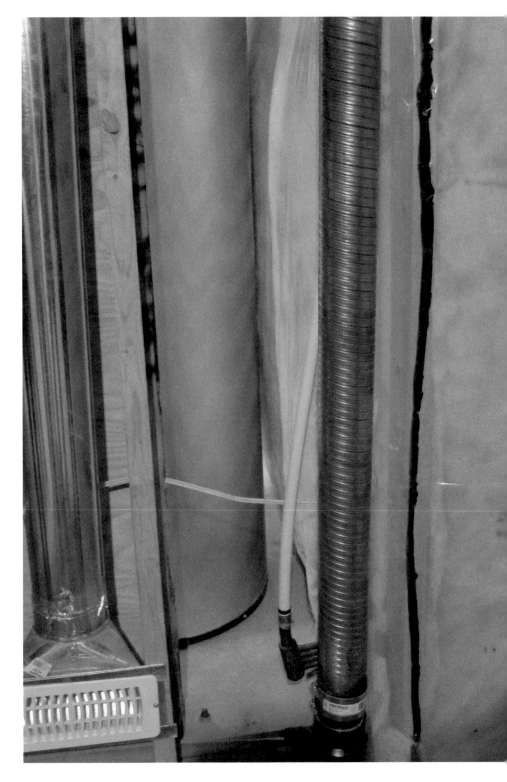

5.28
A built-in system for heat
recovery of waste water heat.

5.29
The Active House has good
daylight conditions.

Peder Vejsig Pedersen

GREAT GULF ACTIVE HOUSE – First Active House in Canada

Location:	Throrold, Niagara Region, Ontario, Canada
Realised:	November 2013
Area:	Lot area: 600 m^2
	Ground floor: 153 m^2
	Upper floor: 154 m^2
Builder:	Great Gulf
Architect:	Toronto architecture firm superkül and a team of Danish architects
Engineer:	Enermodal Engineering, a member of MMM Group

The Great Gulf Active House by the large Canadian developer company Great Gulf Homes was built and presented to the public in November 2013 using prefabricated constructions from Toronto-based Brockport Home Systems.

With its focus on the three quality areas of active houses: comfort, environment and energy, the Great Gulf Active House looks very different than normal houses in Canada, especially with respect to overall architecture and the optimised access to daylight and comfort. (www.activehouse.info)

Concerning energy use there is a much-improved insulation combined with three-layer energy windows with sun shading and both remote controlled natural ventilation as well as heat recovery of the ventilation air, water heat recovery, use of solar collectors for domestic hot water and use of electricity-saving LED lighting and home automation system. Environmentally rainwater is reused for toilet flush, there is used upgraded low-flow plumbing fixtures and an Eco Paver Permeable Driveway and both gas and electricity are supplied from renewable energy sources.

5.30
Brochure information on the great Gulf Active House from 2013.

GREAT GULF ACTIVE HOUSE – First Active House in Canada

Similar to the latest smartphone, the concepts contained within the Great Gulf Active House have a huge potential to trigger an emotional appreciation for technological innovation and design strategies that make for healthy and sustainable living.

5.31
View of Great Gulf Active House from the brochure. The only renewable energy is solar thermal collectors for domestic hot water, but electricity in the area is coming from hydroelectric production from the Niagara Falls area.

GREAT GULF ACTIVE HOUSE – First Active House in Canada

5.32
Installations and prefabricated construction.

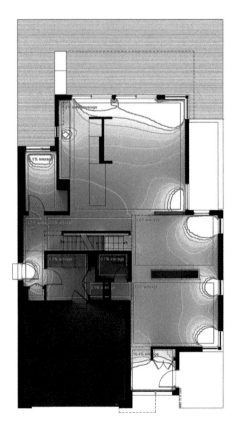

5.33
A daylight calculation and completed radar diagram concerning energy, indoor climate and sustainability compared to a completed radar diagram for a conventional housing unit.

Peder Vejsig Pedersen

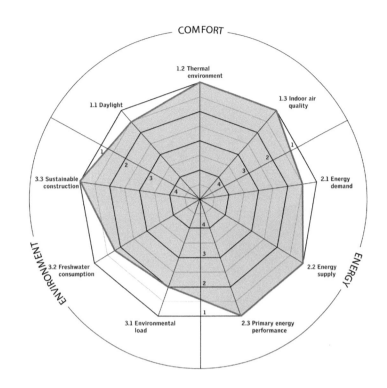

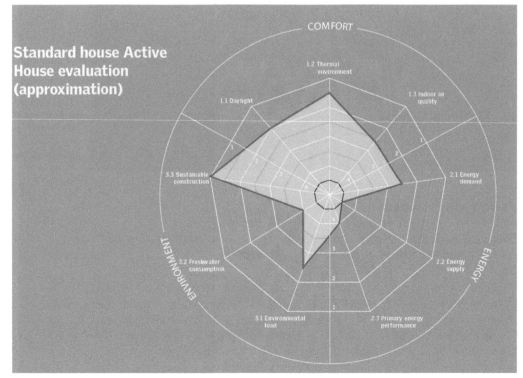

Standard house Active House evaluation (approximation)

5.34
The Canadian Active House developer Great Gulf Homes also makes 70-storey apartment blocks in Toronto. How sustainable this solution is can be discussed.

Performance documentation

Swedish experiences regarding calculated and measured energy consumption in newer low-energy buildings

Performance documentation will be a must by 2020 according to the EU Building Directive. In Sweden it is already implemented. Sweden has already decided to verify the energy quality of new building projects. Within two years at least one year of measurements is necessary as a basis for a final certification of the construction by an energy expert. The two years correspond with the normal warranty period for new constructions. The developer is responsible for the verification and should also take practical use into account, e.g. with focus on the indoor temperatures and actual hot water consumption

The new requirements mean that both building owners and contractors adopt a new practice of measuring heat, hot water and electricity consumption in new buildings. It also means that interesting information about energy consumption in modern low-energy buildings is gathered.

Kommandören and Flagghuset

The Engineering Consultant Company WSP has carried out measurements in the settlements of Kommandören and Flagghuset in Western Harbour of Malmö, where the buildings were built between 2007 and 2009 (see also www.ek-skane. se/infomaterial/rapporter). The expected use was of a maximum of 120 kWh/m^2 per year, but was exceeded by 60–70 per cent when measured; the power requirement was 25–30 W/m^2, against the passive house requirement of 10 W/m^2.

WSP has found that it is important to develop good measurement systems. As an example, it was difficult to get decentralised electricity consumption for ventilation measured. Conditions that caused problems were user behaviour, operational staff competence and commissioning of engineering. It was concluded that monitoring and measurement should be included in the design in the future.

MKB Fastigheter in Malmø

MKB Fastigheter in Malmø has 72,000 apartments. One of the lessons MKB has learnt is that all tenants who have ventilation with heat recovery are extremely pleased with this, and that two layers of glass in the windows is not acceptable. Lessons from MKB are collected in the book *MKB New Function Claims for Production*.

MKB is able to monitor indoor temperatures via Wi-Fi in almost all apartments. At the same time, MKB has introduced an energy-monitoring system that can monitor a wide range of energy information. In several cases it has tackled problems. As an example, there was a tenant in a passive house dwelling who had a very high indoor temperature. Because of poor information the heating system had been programmed to try to keep the temperature at 33 °C in the home. In another case the apartment was very cold in the middle of March (17–18 °C). It was only after six months of investigation that someone who came to look at the supply temperature for district heating could see that it was not put back on after repair. These examples show how important it is to collect and use data from these properties.

Another lesson is that user training is also important. For example, it has been found that if a window is always ajar, it can strain heating consumption by an additional 25 per cent (25 kWh/m^2 per year).

The experience is also that one cannot rely on vendor information as it is usually too optimistic. It is necessary to make a reliability control. This is particularly true for ventilation.

Conclusion

The new requirements for verification of energy quality with measurements focus on key opportunities for improvement in the construction sector, because the measurements can reveal important functional problems. In this context it would be advantageous to repeat the energy calculations in each phase of the project as a natural follow-up (pre-project, planning and realisation). When contractors are involved in verifying energy results, they will be keen to avoid the problems described in this chapter. One idea would be to let the turn-key contractor be responsible for operation in the early years of the life of the building.

Swedish experiences of demands on documentation of energy-saving results in practice

At a conference held in Malmö on 7 April 2011 there was an interesting presentation by the authority 'Boverket', which is responsible for the building rules in Sweden. Focus was on the new Swedish building policy, which includes a demand for verification by monitoring the energy quality of new building projects. This means that within a two-year period after the realisation of a new building project, it is necessary to have at least one year of monitoring results by an independent energy expert in order to make a final certification of the building project.

The builder is responsible for the verification and it is also recommended to take into account the practical use of the building, e.g. with respect to influence of indoor temperatures and hot water use in practice.

The new demands have led to almost a revolution in the attitude to practical energy savings in Sweden, so large building contractor companies are now in close contact with the builders implementing means for practical monitoring and documentation concerning use of heating, domestic hot water and electricity in new building projects.

At the same time, a new standard concerning energy in building projects, 'SVEBY', has been developed.

This consists of three main elements:

- calculations (based on standardised energy data);
- agreement (energy demands and ways to handle non-performance concerning these);
- verification (method for monitoring, follow-up and analysis)

It is still being discussed here whether the above-mentioned should be combined with a wider use of guaranteed key performance demands concerning individual technical solutions in the building projects to reach the energy-saving targets in practice.

An important reason for the practice in Sweden is experiences with much higher energy use (60–70 per cent) in several new building projects compared to what was calculated. One way to secure better results in practice is to insist on making energy calculations in all building phases, including pre-design, design and realisation. When the contractor is made responsible for the results, they will also support this. It will be important to use this practice also for large renovation projects.

Active roofs and facades in sustainable renovation

The Nordic Built 'Active Roofs and Facades' project has been supported by Nordic Innovation to develop leading Nordic competences in the building renovation area in a strong way by creating transnational public–private partnership models to support the development towards nearly zero-energy building solutions and associated performance documentation which is also demanded in the EU Building Directive (see www.europeangreencities.com).

The proposed cooperation with the building industry on developing models and demonstration of Active House-based sustainable renovation is foreseen to create a strong Nordic cooperation in this area.

The project runs from 2014 to 2017 and will involve companies which are represented in the Nordic countries and companies from the international Active House Alliance, whereby the development will use the best transnational competences and networks, creating large possibilities for export of technologies.

The background of the situation in both Nordic and European building renovation projects, where actual energy use is quite often 30–40 per cent higher in practice compared to what was expected from the calculations, and where innovative solutions are only seldom used, is very much connected to the way the building industry is organised. Here, consultants will normally only want to operate in a conservative way, because they are not only selling their expertise, but also the insurance that goes with it, and also because consultants' fees have been reduced considerably, so it is common to work with well-known large suppliers who can contribute to large parts of the design process. This means there is a clear tendency, e.g. not to choose the most energy-efficient solutions, but to allow more mediocre and old-fashioned solutions that the suppliers prefer. When it is common knowledge that detailed performance of equipment in practice is never controlled, there is no incentive to perform better, and higher energy use will often be explained by user behaviour.

In the proposed Nordic Built project a main issue will be to realise the involved renovation projects in a much better way and secure a positive involvement of consultants, so they can be more proactive, e.g. by full-scale testing of innovative solutions before large-scale implementation and by monitoring of key performance indicators also as a basis for negotiating guarantees for performance results as part of the overall procurement process, something which also might be used to avoid normal tendering in connection to development of renovation projects.

An important challenge is to introduce holistic-oriented demands in the so-called Nordic Built Charter in practice in involved demonstration projects.

NORDIC BUILT

We, the Nordic building sector, will join forces and capitalise on our common strengths to deliver the sustainable solutions the Nordic region and the world demands. The time is now and the principles of Nordic Built Charter will lead the way.

OUR COMMITMENT

We commit to taking leadership and implement the Nordic Built principles in our work and our business plans. We commit to taking the necessary actions to deliver competitive concepts for a sustainable built environment that benefit users, the building sector, our region and the world.

OUR NORDIC BUILT PRINCIPLES

WE WILL CREATE A BUILT ENVIRONMENT THAT:

Is made for people and promotes **O1** quality of life	**O6** Is robust, durable, flexible and timeless - built to last
Pushes the limits of sustainable **O2** performance, as a result of our innovative mind-set and high level of knowledge	**O7** Utilises local resources and is adapted to local conditions
Merges urban living with the **O3** qualities of nature	**O8** Is produced and maintained through partnerships founded on transparent collaboration across borders and disciplines.
Achieves zero emissions **O4** over its lifecycle	**O9** Employs concepts that are scalable and used globally
Is functional, smart and aesthetically **O5** appealing, building on the best of the Nordic design tradition	**10** Profits people, business and the environment

OUR INVITATION

We, the Nordic building sector, invite the Nordic governments and public authorities, investors and financial institutions, end-users and building owners, the energy sector and all others who have a stake in our mission, to join us in our efforts to accelerate the transition to a sustainable built environment.

Signed by:

Company name	Company representative	Date	Signature

norden
Nordic Innovation

5.35
Overview of the Nordic Built Charter, which tries to set a sustainable building approach for building and renovation with a focus on people, business and the environment.

Ellebo Garden Room

The winner of the Danish Nordic Built Challenge international architectural competition, Ellebo Garden Room by Adam Kahn Architects from London, is targeted for realisation in Ballerup, west of Copenhagen, from 2015. Financing by the Danish Social Housing Fund makes a high-quality solution possible.

5.36
The winning renovation design by Adam Kahn Architects from London.

5.37
Interior of an apartment.

5.38
Ellebo today.

There are examples of projects where facade or roof elements with integrated features have been used for renovation, e.g. in Austria, Germany, Denmark and Norway. In these projects the solutions have been developed case by case, and no general solution exists. As the existing buildings are always individual, the specific characteristics will have to be taken into account in any case, but the project team is confident that an ideal solution can be developed, which will integrate the most important features required in typical renovation projects. The project will develop a concept that will fulfil the typical renovation needs of the Nordic buildings that are most often in need of renovation.

Cenergia has worked with Kuben Management, Copenhagen municipality and other partners on the large EU Concerto Green Solar Cities project with demonstration of low-energy renovation and use of solar energy in Copenhagen and Salzburg (www.greensolarcities.com).

Cenergia and other partners from Denmark have since 2003 been involved in a cooperation with large Nordic companies like the Velux group to develop zero-energy building concepts with a special view to renovation, and has here been involved in development of best-practice technologies like the SOLTAG CO_2-neutral rooftop apartment from 2005 (www.soltag.net) and SOLARPRISM from 2009 and 2012 (www.youtube.com/watch?v=RxgRDXaMH-M), the last-mentioned as part of developing the so-called 'Albertslund Concept', which aimed at sustainable and energy-efficient renovation of 4,000–5,000 housing units in Albertslund, west of Copenhagen. In 2012 this led to Albertslund being elected as Nordic Energy Municipality 2013 (www.albertslund.dk – Politik – Politikker og Strategier – Klimaplan – Albertslundkonceptet). The solar solution has been further developed and has been used for renovating ten social houses in Montfort, the Netherlands, moving those buildings from an energy classification of E to A++ according to the Dutch scheme. Experiences from the Plan C project on Sustainable and Energy Efficient Renovation (www.plan-c.dk) coordinated by Gate21 in 2010–2013 with a total budget of DKK40 million will also be utilised.

VTT from Finland has been involved in the development of the facade element and the renovation of an apartment building in Riihimäki, Finland. In this project windows, high-level insulation and inlet air ducts were integrated in the timber element facade.

Many of the buildings built in large volumes in the 1970s are or will soon be in need of renovation. In most Nordic countries these buildings are built with similar technology, and therefore the solutions that can be applied in a typical building from that era have great replicability potential. The Nordic cities are often owners of several similar buildings.

The Active Roofs and Facades project will be able to develop leading Nordic competences in the building renovation area in a strong way by creating trans-national public–private partnership models to support the development towards nearly zero-energy building solutions and associated performance documentation which is also demanded in the EU Building Directive. The proposed cooperation with the building industry on developing models and demonstration of Active House-based sustainable renovation is foreseen to create a strong Nordic cooperation in this area, which at the same time will have a large export potential.

The aim is to cooperate with both the City of Copenhagen as well as other local stakeholders on the further development of the concepts developed in the EU Concerto Green Solar Cities project, including use of prefabricated and system-based BIPV solutions, use of innovative and low-cost building-integrated and decentralised HRV systems, use of advanced energy supply and smart grid solutions, together with intelligent systems for survey and follow-up of energy and comfort results.

In connection to social housing renovation there will be in the project a cooperation with the housing association KAB from Denmark and with the housing developer Trianon, working with WSP from Sweden, both of which are involved in the Nordic Built Challenge.

Besides Active House qualities there will be a main focus on socio-economic activities, e.g. how can you do housing renovation and keep the tenants in the apartments to avoid extra costs for rehousing? There will also be education of tenants to allow them to utilise the sustainable renovated housing units in the right way.

The basic idea of the Active Roofs and Facades proposal is to develop solutions that can be integrated into buildings and support a completely new way to renovate existing buildings, which is based on:

- high-quality insulation without thermal bridges;
- new active facades with new glazing design and multifunctional windows;
- use of decentralised HRV design integrated into the building envelope and windows;
- use of BIPV on the roof and façade;
- being part of an overall smart grid approach for electricity and district heating;
- use of integrated daylight solutions, optimising the daylight in buildings and reducing energy for lighting;
- use of integrated performance documentation focusing on all the require-ments in the Nordic Build Charter;
- use of the Active House Specifications to document the performance of the solutions. See the Active House initiative at www.activehouse.info.

At present it is the idea to develop a number of apartment building projects as well as school projects in the Nordic countries and renovate them with the use of active roofs and facades. The renovation follows the ten parameters in the Nordic Built Charter as well as the Active Roofs and Facades workplan. The performance of the renovation will be documented using the Active House Specification and evaluation.

A Nordic innovation initiative for development of active roofs and facades as part of sustainable energy-efficient renovation with documented performance

1. There will be utilised an overall Active House design approach.
- Introduction of Active House integrated energy and environmental design.

- This includes confirmation of key performance indicators in full-scale test rooms before implementation of demonstration projects.

2. There will be a focus on optimised energy supply, including advanced low-temperature heating systems, use of renewable energy and active roofs and facades.
- Investigation of innovative solar energy combined heat and power systems.
- Building integrated PV and solar thermal as part of the overall innovative active roofs and facades configuration concept.
- Advanced district heating design based on low-temperature operation and possibility for summer halting of operation to avoid district heating losses in periods with low consumption.
- Investigation of innovative PV thermal solutions.

3. There will be developed, tested and demonstrated an optimised building envelope including decentralised ventilation and daylight.
- Advanced and innovative prefabricated insulation solutions, according to a systemic approach voiding cold bridges and air leakage based on passive house experience.
- Utilisation of building mass.
- Advanced and innovative window solutions including solar shading and day-lighting function.
- Decentralised heat recovery and no-noise intelligent ventilation design integrated in building and building envelope in correspondence with PV-assisted natural ventilation and removal of toxic particles by advanced filters.

4. According to the Active House principles, there will be an Integrated Performance Documentation System.
- Use of Active House specifications in connection to agreed 'performance verification procedure', including use of energy signatures.
- Access to main electricity use and heating and renewable energy data through the internet, including lighting, ventilation, appliances, district heating, solar heating, PV electricity.
- Building and district energy data through a building energy management information system.
- Control of possible problems with moisture.
- Control of indoor air quality.

5. It is the idea to use an advanced lighting system.
- Use of LED technology.
- Utilisation of daylight.
- Optimised user control.

6. There will be an overall smart building and solar energy approach with a focus on guaranteed energy-saving results and lifetime optimised economy.
- Control of major key performance indicators.
- Continuous survey of energy performance, e.g. via Apps.

Chapter 6

Smart Solar Cities

PEDER VEJSIG PEDERSEN

It is very relevant to take on the challenge of working with the 'smart cities' of the future by including a strong plan for implementation of solar energy, which makes it possible to widen the concept into 'smart solar cities'. This is a strong vision when you take the extremely high reductions in costs into account, both for solar PV technologies and for solar thermal. Recent PV cost reductions actually mean that PV systems can prove a better economy today than off-shore wind turbines.

There is also strong support for the practical implementation of solar energy technologies in relation to the much-increased low-energy building demands in Building Regulation standards both for new-build and large renovation projects.

In Denmark the most advanced low-energy building standard is the low-energy class 2020 standard, which is the Danish definition of a 'nearly zero-energy' building standard. This is approximately 25 per cent better than the low-energy class 2015 standard, which is already a demand in many Danish cities, and the general attitude is that the increased low-energy quality from the low-energy class 2015 to 2020 will to a large extent come from local use of renewable energy solutions, and here the use of solar energy is the most obvious solution.

The green solar cities vision is based on the universal relation between the necessary initiatives you need to work with for the future. That means energy savings in both new-build and renovation are the first thing you need to introduce and optimise. When this is done you should look first to an optimised energy supply solution and second investigate how solar energy can be utilised with a high contribution in connection to this, e.g. making nearly zero-energy or plus-energy building possible. The lack of focus on performance documentation in practice is still an important barrier here.

In connection to the finalisation of the Green Solar Cities Concerto project in Copenhagen and Salzburg, the situation is that important steps have been taken but it can be concluded that solar energy solutions in buildings are still in most cases an exception. For Copenhagen a positive development has been seen, but there is still reluctance towards it from decision-makers – important architectural and engineering companies talk about how they want to avoid these solutions. In Copenhagen the municipality is positive concerning solar PV, but only sees it as a very small contribution to CO_2 reduction. They prefer to avoid solar thermal because their district heating company has the old-fashioned attitude that it does not fit well into combined heat and power. In Salzburg they have not really taken up solar PV yet, but they are very strong with respect to implementing solar thermal systems and they integrate them both in district heating and in relation to combined heat and power, due in particular to their many years of use of a special energy-point financial system for social housing.

However, for both cities solar energy solutions are still not mainstreamed yet. When large building projects are started, both architects and builders will normally not integrate solar solutions from the beginning, but rather see it as an add-on, e.g. to fill gaps in reaching a Danish low-energy class 2020 standard, which you have to live up to in Copenhagen in relation to municipal buildings or new urban development areas. This is a policy which it is hoped will be maintained, even though a new proposal for changing the Danish Planning Law has recently been proposed.

Due to having the highest energy costs in Europe since the early 1980s (except for industrial organisations), Denmark has a unique opportunity to be a frontrunner in the necessary transformation to a renewable-energy-based society. The result of this policy, where the politicians decided to keep the costs of oil, gas and electricity high via taxation as a result of the second oil crisis in the early 1980s has led to a situation where Denmark has a clear leadership in the use of district heating, with 55 per cent of buildings utilising this, often together with combined heat and power as well as wind energy, which is now covering around 30 per cent of yearly electricity production.

Another not so recognised result of the situation with high energy prices is that solar energy solutions have started to penetrate the energy market on a large scale. During the last ten years we have seen a large number of solar thermal plants being connected to the widespread district heating systems in Denmark, typically covering up to 20 per cent of the yearly district heating demand and without any kind of subsidies, and in some cases covering up to 50 per cent of the yearly district heating demand when coupled with large seasonal storage solutions.

For the PV electricity market, which has been kick-started by German feed-in tariffs from around 2000, there was not much happening in Denmark based on a yearly net-metering scheme (which was also used for the wind turbines in the beginning), but when huge reductions in PV costs emerged in the last few years there was suddenly a situation in which PV systems could be installed with good economy for the users, something which especially one-family house owners became aware of.

The result was that where there was an almost non-existent market in 2010, a small market was established in 2011 with 12 MWp, and this rocketed to approximately 500 MWp PV installations in 2012. This is something that had to be handled politically in an energy system like the Danish one, because energy taxes are a big part of the state budget every year. Very quickly complaints were made by Danish Energy, a cooperative of energy-producing companies, about one-family house owners who invested in the allowed up to 6 kWp PV installations, and in fact could avoid any electricity bills for the next 30 years, even though

Peder Vejsig Pedersen

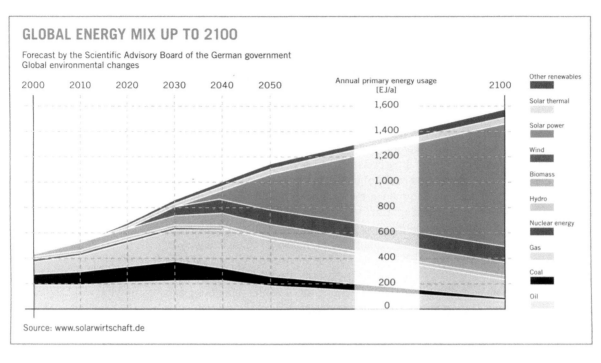

GLOBAL ENERGY MIX UP TO 2100

Forecast by the Scientific Advisory Board of the German government
Global environmental changes

| 2000 | 2010 | 2020 | 2030 | 2040 | 2050 | | 2100 |

Annual primary energy usage
[EJ/a]

1,600
1,400
1,200
1,000
800
600
400
200
0

Other renewables
Solar thermal
Solar power
Wind
Biomass
Hydro
Nuclear energy
Gas
Coal
Oil

Source: www.solarwirtschaft.de

6.1
Forecast for global energy mix by the Scientific Advisory Board of the German government. Here it is clear that solar power is expected to be the main energy source in the future. This means that implementation strategies are needed today for future-oriented frontrunners.

they needed to utilise the common electricity network, especially during winter when there is not so much sun in Denmark.

The politicians were unaware of this situation in early 2012 when they signed a long-term energy strategy aiming at transforming Danish society into a 100 per cent renewable energy supply society, mainly based on off-shore wind energy, by 2035. Here they were only allowing 800 MWp solar PV to be installed by 2020. The situation now is that the yearly net-metering scheme for solar PV has been changed into an hourly based net-metering scheme as part of a smart grid strategy and future PV installations are aimed to have a stronger focus towards housing associations and cities.

With the continuous reduction of installed PV costs, the situation in Denmark today is that PV systems in many cases are becoming more cost-effective than off-shore wind installations. This situation is still not recognised by politicians, but it creates the opportunity for a basis for large-scale implementation of solar energy solutions in Denmark.

Here it is relevant to look at the situation we have in Denmark, where it is being discussed what to do with the CHP-based district heating, when the main source of electricity is wind turbines. However, large-scale use of solar energy provides an opportunity for a win–win situation if it is installed in the form of so-called solar energy combined heat and power, which provides solar heating and solar electricity in a 1:1 ratio, in the same way as the existing large-scale CHP system does. This means operation of the large CHP system can be reduced in sunny periods, saving fuel as a result. Here the challenge is to reach low costs for the solar thermal installations. This has already proven to be possible for large ground-mounted solar thermal plants at total installed costs around €200 per square metre, while it has been difficult to reach this cost level for solar thermal collectors placed on buildings, which can cost 2–4 times more.

In the Green Solar Cities partner city, Salzburg, the solar thermal collectors are connected directly to the district heating system, making lower costs possible, while solar thermal collectors for domestic hot water systems, as used in Copenhagen, can cost four times more than ground-mounted solutions. Bringing solar thermal collector costs down is thus a major challenge for the future.

However, a main obstacle is still the lack of integrated performance documentation, even though continuously increased demands concerning energy efficiency, like the low-energy classes 2015 and 2020 in Denmark, are already utilised by many cities. The result of this is that builders cannot be certain of obtaining the high energy performance that has been paid for, since the only documentation relies on calculations. Until today we have only experienced one area where a much-increased energy quality has been introduced in practice in Denmark, and that is the demand for airtightness of buildings, since it is easy to control and there are still responsibilities for contractors up to five years after a building is completed.

Here it is a clear challenge to introduce much more useful monitoring and survey systems so it is possible in a short time to establish what the performance of a new building is in practice, so this quality can be included in the way buildings are realised. This should also be the approach for building renovation.

In connection to this it is very important not only to focus on how to make new building projects, but to try to utilise experiences from already realised solar low-energy building projects, and here especially not only to focus on the good results but also try to learn from the not-so-good results, so the same failures are not implemented again and again.

As support for a smart cities development it is suggested that cooperation with social housing organisations in Denmark could help realise what you could call smart grid housing renovation. In Denmark that can be considered to be very relevant because large investments are made in housing renovation every year. The basis for this is that it is common to aim for 'climate shield' renovation for social housing in Denmark, in most cases with financing from the social housing fund. Other improvements – like installations – are normally more difficult to finance.

In addition to improvement of the climate shield with improved insulation and new windows, the use of heat recovery ventilation is considered to be an important option for improved energy efficiency and improved indoor air climate. A barrier is the cost, which in most cases will amount to €6,000–7,000 per apartment. Here, it can be a solution to utilise a new technology with window-integrated heat recovery ventilation and utilisation of automatic filter exchange boxes, which have prospects of being able to be realised at around half of the cost.

To be able to reach a low-energy class 2020 standard it is suggested to combine this approach with the use of solar energy combined heat and power, which is a good match to CHP-based district heating. It is the idea to utilise the new 'smart grid' oriented legislation concerning PV systems in Denmark. The fact is that you will get a good price for the produced solar electricity if you consume it in the same hour as it is produced. For social housing apartments the result of this is that it will in most cases be best to limit the PV system size per apartment

Peder Vejsig Pedersen

to 1.5 kWp, which can produce 1,200–1,300 kWh of electricity on a yearly basis. If you combine this with 2–3 m^2 of thermal solar collectors per apartment for domestic hot water, then you both get a combined solar energy supply, which matches the CHP-based district heating in a good way, and you also secure a situation in which you can discontinue the connection to the district heating system in the summer months, in this way securing reduced distribution heat losses.

This approach has already been demonstrated in the EU Concerto Green Solar Cities project in Valby in Copenhagen, but only with some of the mentioned features, since the possibility to avoid connection to the district heating in summer was not a feature. Also, there was no aim to realise integrated solutions for the solar energy roofs, which is a very interesting challenge for the future.

It is proposed at the same time to link the initiative with a performance documentation approach, e.g. in connection to the 'Active House' standard (www.activehouse.info), and to focus on prospects for an improved lifetime of PV panels, e.g. with use of new types of glass/glass PV modules and also to consider use of local electricity use management and electrical storage options as well as part of a 'smart grid' strategy.

In several European countries and cities there is huge interest in being involved in a 'smart city' development; at the same time plans to become CO_2-neutral during the coming years also have strong support, often linked into local climate plans like the Climate Plan of the City of Copenhagen to be CO_2-neutral by 2025, thereby reducing CO_2 emissions by 1.16 million tons (see also www.cphcleantech.com and www.kk.dk). At the same time there is a vision of supporting what is often called green growth, because the transformation of society to rely 100 per cent on renewable energy and even be CO_2-neutral is seen as a strong business, which can help to create the businesses of the future.

In Salzburg in Austria there is support for this development, e.g. in the 'Smart Grids Model Region Salzburg' (www.smartgridssalzburg.at), and since 2012 a new housing area utilising smart grids has been developed (www.rosazukunft.at); in another area a low-voltage grid is combined with use of PV systems and electric cars.

When we look at Denmark, there are also many cities working on smart city development and climate plans to be CO_2-neutral.

A strong initiative that can be mentioned is Project Zero (http://brightgreen business.com), aiming to make the Sønderborg Region, with 80,000 people in the south part of Jutland, CO_2-neutral by 2029. Typically for Denmark, the smart grid approach focuses on the electrical grids as well as the large district heating grid, which will be extended and utilise a mix of renewable energy solutions, like solar thermal (4 per cent), biomass, biogas, geothermal, waste and heat pumps.

For electricity, 80 per cent is aimed to come from wind turbines, 4 per cent from PV and 16 per cent from waste and biogas. There will be 113 MW of off-shore wind turbines, which will provide 407 GWh/year, equal to 23 per cent of future CO_2 emissions in 2029 (estimated investment in 2009 of €282 million). This could also be covered by 448 MWp PV as an alternative. For the countryside 65 per cent of heating demands will be covered by heat pumps, 15 per cent from solar heat and 20 per cent from biomass. The city of Skive in the middle of Jutland also has a strong climate plan, aiming at being CO_2-neutral by 2029.

The way to become CO_2-neutral in Skive by 2029 is to go from the existing total energy consumption of 6,500TJ in 2010 to 8,800TJ in 2029, where only 2,000TJ is still fossil fuels. To match this an extra amount of renewable energy will be produced for export outside of the community (especially wind power and geothermal energy). Here, renewable energy sources are providing electricity and heat to create the necessary balance in the overall energy system.

Particularly strong elements of the Energy Action Plan in Skive are:

- Photo-Skive. The largest municipal implementation in Denmark of PV systems with 1.5 MWp has been implemented in Skive since 2009. This means that all municipal buildings had PV systems by 2013.
- Expansion of wind turbine capacity. By 2007, 85 MW of wind turbines were in use in Skive; a new capacity of 45 MW will be introduced by 2029, of which the municipality is responsible for 30 MW.
- A biomass gasification system will provide CHP for Skive district heating company.
- An electric car strategy will lead to 450 TJ of savings in 2029, including electric cars for the municipality. Renewable energy electricity will provide for this.
- Electricity savings in municipal buildings will be promoted by use of a digital tool.
- 40 per cent of the existing natural gas consumption will in a few years be transferred into 'green gas' (biogas and H_2) introduced in the existing natural gas network. Skive is leading the EU-Intereg-supported 'Implement' project in cooperation with Samsø, Lemvig, Norway and southeast Sweden. There will be 580TJ from biogas by 2029.
- Skive has been the major promoter of solar heating in Denmark since 1994, where the first 375 m^2 solar heating installation was made at a school. Now there are 55 solar heating systems with 50,000 m^2 for all municipal buildings. A large solar heating park of 50,000 m^2 is being prepared in Skive. This could be coupled with a similar energy amount of PV electricity as an innovative solar energy combined heat and power approach. By 2029 solar heating will go from 707TJ to 2,450TJ.
- Geothermal energy is now being introduced in west Salling in the Skive municipality and this is being combined with an expansion of the existing district heating systems in five areas of Skive, so they are coupled together as one large district heating network.
- The village of Durup will become CO_2-neutral in 2015.
- The city hall will become CO_2-neutral.
- The island of Fur will be made into a model of sustainable energy.
- 'My new home party' (inspired by Tupperware) is used to promote energy renovation of all one-family houses, reaching 5,000 housing units.
- Energy check, analysis and project support for building owners.
- All new buildings will live up to a one step higher energy quality than is demanded in the Building Regulations. Today this means the building class 2015.
- Buses running on biogas will be introduced.

Smart Cities development in general also includes ideas of making the citizens into 'prosumers', where they function both as consumers of energy as well as suppliers of electricity back to the electricity grid.

Also the 'Bright Green Island' project, on the island of Bornholm, is being developed into a test site for new green technologies and smart city development.

Photographer list

Active House	4.120, 5.15, 5.16, 5.17, 5.18, 5.19, 5.20, 5.21, 5.22, 5.23, 5.24, 5.25, 5.26, 5.27, 5.28, 5.29, 5.30, 5.31, 5.32, 5.33, 5.34
Anita Jørgensen	2.4
Arkitektfirmaet Ole Dreyer	2.12, 2.13
Carl Galster EUDP	4.135
Cenergia	1.28, 1.41, 1.44, 1.45, 1.46, 1.49, 1.50, 1.51, 1.53, 3.6, 3.7, 3.13, 3.14, 3.27, 3.33, 3.34, 3.35, 3.36, 3.42, 3.46, 3.47, 3.53, 3.62, 4.23, 4.96, 4.97, 4.98, 4.99, 4.100, 4.102, 4.111, 4.114, 4.115, 4.118, 4.119, 4.121, 4.122, 4.142, 4.143, 4.145, 4.146, 4.157, 4.199
Copenhagen Municipality	2.21
Domea 2005	4.117
Dorthe Krogh	2.17, 2.18
Ecovent	4.128
Ecovent/Øland brochure	4.158
Energimidt	2.24, 2.26
Entasis Arkitekter	2.10
Erik Møller Arkitekter	2.15
FSB 1999	4.49, 4.53
Gaia Solar	2.20, 2.22
German Solar	2.16
Inge Straßl	1.3, 1.4, 1.5
Jakob Klint	1.1, 1.2, 1.8, 1.12, 1.71, 1.72, 1.76, 3.22, 3.23, 3.52, 3.58, 3.59
Jens Lindhe	2.1
Johan Galster	2.14
Jyllandsposten 2003	4.174
KAB 1992	4.44
KAB EU partner	4.153
Karin Kappel	2.2, 2.3, 2.6, 2.7, 2.8, 2.11
Klaus Boyer Rasmussen	1.32
Komproment brochure	4.195
Lin Kappel	2.5
Malmø brochure	4.211
Maria Wall (partner)	4.1, 4.2, 4.3
Martin Rubow	1.29, 1.61
Martin Rubow/EUDP	4.130
NBHA 1995,	4.94, 4.95

Nordic Build website	5.35, 5.36, 5.37, 5.38
O Langenkamp	4.6
Peder Vejsig Pedersen	1.7, 1.9, 1.10, 1.11, 1.15, 1.16, 1.18, 1.19, 1.20, 1.30, 1.31, 1.33, 1.34, 1.35, 1.36, 1.37, 1.47, 1.48, 1.52, 1.54, 1.55, 1.56, 1.57, 1.58, 1.68, 1.69, 1.70, 1.75, 3.1, 3.2, 3.3, 3.4, 3.5, 3.9, 3.10, 3.15, 3.16, 3.17, 3.18, 3.19, 3.20, 3.21, 3.25, 3.26, 3.28, 3.29, 3.30, 3.31, 3.32, 3.37, 3.39, 3.40, 3.41, 3.50, 3.51, 3.54, 3.55, 3.56, 3.57, 3.60, 3.61, 3.64, 3.65, 3.66, 3.101, 3.102, 3.103, 3.104, 3.105, 3.106, 3.107, 3.108, 3.109, 3.110, 4.5, 4.7, 4.8, 4.9, 4.10, 4.11, 4.12, 4.13, 4.14, 4.15, 4.16, 4.17, 4.18, 4.19, 4.20, 4.21, 4.22, 4.24, 4.25, 4.26, 4.27, 4.28, 4.29, 4.30, 4.31, 4.32, 4.33, 4.34, 4.35, 4.36, 4.37, 4.38, 4.39, 4.40, 4.41, 4.42, 4.43, 4.45, 4.46, 4.47, 4.48, 4.50, 4.51, 4.52, 4.54, 4.55, 4.56, 4.57, 4.58, 4.59, 4.60, 4.61, 4.62, 4.63, 4.64, 4.65, 4.66, 4.67, 4.68, 4.69, 4.70, 4.71, 4.72, 4.73, 4.74, 4.75, 4.76, 4.77, 4.78, 4.79, 4.80, 4.81, 4.82, 4.83, 4.84, 4.85, 4.86, 4.87, 4.88, 4.89, 4.90, 4.91, 4.92, 4.93, 4.101, 4.103, 4.104, 4.106, 4.107, 4.108, 4.109, 4.110, 4.112, 4.116, 4.125, 4.126, 4.127, 4.132, 4.133, 4.134, 4.136, 4.137, 4.138, 4.139, 4.140, 4.141, 4.144, 4.147, 4.148, 4.149, 4.150, 4.151, 4.152, 4.155, 4.159, 4.160, 4.161, 4.162, 4.163, 4.164, 4.165, 4.166, 4.167, 4.168, 4.169, 4.170, 4.171, 4.172, 4.173, 4.175, 4.176, 4.177, 4.178, 4.179, 4.180, 4.181, 4.182, 4.183, 4.184, 4.185, 4.186, 4.187, 4.188, 4.189, 4.190, 4.191, 4.192, 4.193, 4.194, 4.196, 4.197, 4.198, 4.200, 4.201, 4.202, 4.203, 4.204, 4.205, 4.206, 4.207, 4.208, 4.209, 4.210, 4.212, 4.213, 4.214, 4.215, 4.216, 4.217, 4.218, 4.219, 4.220, 4.221, 4.222, 5.1, 5.2, 5.3, 5.4, 5.5, 5.6, 5.7, 5.8, 5.9, 5.10, 5.11, 5.12, 5.13, 5.14
Planenergi	1.42
Pvexchange	1.60
SIR + Steinbeis	3.67, 3.68, 3.69, 3.70, 3.71, 3.72, 3.73, 3.74, 3.75, 3.76, 3.77, 3.78, 3.79, 3.80, 3.81, 3.82, 3.83, 3.84, 3.85, 3.86, 3.87, 3.88, 3.89, 3.90, 3.91, 3.92, 3.93, 3.94, 3.95, 3.96, 3.97, 3.98, 3.99, 3.100, 4.4, 3.63
Skive Municipality	2.23, 2.25
Svendborg Architects	1.73, 1.74
Velux/EUDP project	4.123, 4.124
VIA UC project partner	4.154, 4.156
WSP project partner	4.223

Bibliography

Janson, U., *Passive Houses in Sweden: From Design to Evaluation of Four Demonstration Projects*. PhD thesis. Lund University, 2010. www.ebd.lth.se/fileadmin/energi_byggnadsdesign/images/Publikationer/Doc_avhandling_UJ_Bok_webb.pdf.

Pedersen, Peder Vejsig, *Solar Energy and Urban Ecology* (in Danish), Nyt Teknisk Forlag, 2002.

Pedersen, Peder Vejsig, *Resurgence EU project: Building Integration of PV*, Cenergia, Common work package, June 2002.

Pedersen, Peder Vejsig, *Solar Low Energy Buildings in Denmark*, Sustainable Buildings South Europe, Turin, 7–8 June 2007.

Pedersen, Peder Vejsig, *Solar Low Energy Retrofit Housing Projects in Denmark*, Sustainable Buildings South Europe, Turin, 7–8 June 2007.

Pedersen, Peder Vejsig, *The First Zero Energy Retrofit Housing Project in Denmark*, Passive House Nordic Countries, Aalborg, 2010.

Pedersen, Peder Vejsig, *Active House Renovation in Albertslund*, 2012.

Pedersen, Peder Vejsig, *Nordic Built Active Roofs and Facades in Sustainable Renovation*, Cenergia, 2014.

Pedersen, Peder Vejsig and Inge Straßl, *Green Solar Cities, EU Concerto Project in Copenhagen and Salzburg*, International Solar Cities Conference, Adelaide, 2010.

Persson, Bengt (ed.), *Sustainable City of Tomorrow: Bo01 – Experiences of a Swedish Housing Exposition*. Formas, 2005.

Straßl, Inge, *Deliverable 2.2: Reward system for ecologic housing in Salzburg. A point-based system for the funding of energy-efficient eco-buildings and the use of biomass and solar energy*. Version 1, 23 May 2008.

Annexes in connection to EU Green Solar Cities project by Peder Vejsig Pedersen

Annex 5: 150,000 m^2 PV modules in Valby (Eurosun 2002 Paper).

Annex 6: Use of solar energy solutions in Denmark.

Annex 7: Electricity use in Valby/Copenhagen.

Annex 8: Prefabricated CO_2 neutral rooftop apartment (also presented at Passive House Nordic Countries, Trondheim, 2008).

Annex 9: PV solutions in Copenhagen.

Annex 12: Balanced mechanical ventilation with counter flow heat recovery and low electricity consumption for fans.

Green Solar Cities reports on GSC website

Danish Association of Sustainable Cities and Buildings, *Best Practice for Low Energy Building*.

Danish Association of Sustainable Cities and Buildings, *Solar and Biomass for Low Energy Building*.

Danish Association of Sustainable Cities and Buildings, *Projects that Transform*

Frendrup, Jens, GSC D7.1 and D7.4, *Reports on Training Results*, European Green Cities, November 2009 and November 2010.

Frendrup, Jens, GSC D2.13, *Final Report on RTD Theme Groups*, European Green Cities, August 2011.

GSC D6.3.1, *Final Implementation Report*, Salzburg, SIR, October 2013.

GSC D6.3.2, *Implementation and Monitoring in Valby*, Cenergia, November 2013 (revised January 2014).

Hasselarr, Evert, GSC D1.6, *User Test of Best Practice RUE and RES Technologies and Recommendations for Improving User Friendliness of HVAC Systems*, OTB Institute, University of Delft, November 2008.

Hasselarr, Evert, GSC D1.14, *Evaluation and Socio Economics*, OTB, University of Delft, May 2013.

Pedersen, Peder Vejsig, GSC D2.1.1, *Catalogue of Best Practice Technologies*, Cenergia, May 2008.

Pedersen, Peder Vejsig, GSC D4.1.1, *The Green Quality Building Process Concepts*, Cenergia, June 2008.

Pedersen, Peder Vejsig, GSC D2.14, *RTD Work and Guidelines on Heat Recovery Ventilation in the EU Concerto Project*, Green Solar Cities, Cenergia, November 2011 (revised 2013).

Rieger, Ursula, Peder Vejsig Pedersen, Vickie Aagesen and Inge Straßl, GSC D6.5, *Final Monitoring Report*, Steinbeiss Institute, Cenergia and SIR, April 2014.

Straßl, Inge, GSC D2.1.2, *Reward System for Ecological Housing in Salzburg*, SIR, May 2008.

Straßl, Inge, Jakob Klint and Jappe Goud, GSC D2.12, *New Financing Models for Sustainable Building*, November 2010.

Websites

www.activehouse.info
www.cenergia.dk
www.danfoss.com
www.ecovent.dk
www.energinet.dk
www.ens.dk
www.europeangreencities.com
www.fbbb.dk
www.gaiasolar.dk
www.greensolarcities.com
www.grundfos.com
www.kk.dk/climate
www.kubenman.dk
www.nordicbuilt.org
www.rockwool.com
www.sir.at
www.solarplan.dk
www.solivalby.dk
www.soltag.net
www.velux.com

Index

Page numbers followed by 'f' refer to figures and followed by 't' refer to tables.